# DOSIMETRY ASPECTS OF EXPOSURE TO RADON AND THORON DAUGHTER PRODUCTS

Report by a Group of Experts
of the OECD Nuclear Energy Agency
September 1983

NUCLEAR ENERGY AGENCY
ORGANISATION FOR ECONOMIC CO-OPERATION AND DEVELOPMENT

Pursuant to article 1 of the Convention signed in Paris on 14th December, 1960, and which came into force on 30th September, 1961, the Organisation for Economic Co-operation and Development (OECD) shall promote policies designed:

- to achieve the highest sustainable economic growth and employment and a rising standard of living in Member countries, while maintaining financial stability, and thus to contribute to the development of the world economy;
- to contribute to sound economic expansion in Member as well as non-member countries in the process of economic development; and
- to contribute to the expansion of world trade on a multilateral, non-discriminatory basis in accordance with international obligations.

The Signatories of the Convention on the OECD are Austria, Belgium, Canada, Denmark, France, the Federal Republic of Germany, Greece, Iceland, Ireland, Italy, Luxembourg, the Netherlands, Norway, Portugal, Spain, Sweden, Switzerland, Turkey, the United Kingdom and the United States. The following countries acceded subsequently to this Convention (the dates are those on which the instruments of accession were deposited): Japan (28th April, 1964), Finland (28th January, 1969), Australia (7th June, 1971) and New Zealand (29th May, 1973).

The Socialist Federal Republic of Yugoslavia takes part in certain work of the OECD (agreement of 28th October, 1961).

*The OECD Nuclear Energy Agency (NEA) was established on 20th April 1972, replacing OECD's European Nuclear Energy Agency (ENEA) on the adhesion of Japan as a full Member.*

*NEA now groups all the European Member countries of OECD and Australia, Canada, Japan, and the United States. The Commission of the European Communities takes part in the work of the Agency.*

*The primary objectives of NEA are to promote co-operation between its Member governments on the safety and regulatory aspects of nuclear development, and on assessing the future role of nuclear energy as a contributor to economic progress.*

*This is achieved by:*

- *encouraging harmonisation of governements' regulatory policies and practices in the nuclear field, with particular reference to the safety of nuclear installations, protection of man against ionising radiation and preservation of the environment, radioactive waste management, and nuclear third party liability and insurance;*
- *keeping under review the technical and economic characteristics of nuclear power growth and of the nuclear fuel cycle, and assessing demand and supply for the different phases of the nuclear fuel cycle and the potential future contribution of nuclear power to overall energy demand;*
- *developing exchanges of scientific and technical information on nuclear energy, particularly through participation in common services;*
- *setting up international research and development programmes and undertakings jointly organised and operated by OECD countries.*

*In these and related tasks, NEA works in close collaboration with the International Atomic Energy Agency in Vienna, with which it has concluded a Co-operation Agreement, as well as with other international organisations in the nuclear field.*

Publié en français sous le titre:

ASPECTS DOSIMÉTRIQUES DE L'EXPOSITION
AUX PRODUITS DE FILIATION DU RADON ET DU THORON

## FOREWORD

This report was prepared by a Group of Experts for the Committee on Radiation Protection and Public Health of the OECD Nuclear Energy Agency. It completes Phase I of the NEA programme of work on radon dosimetry and monitoring and will be followed by Phase II concerned with the practical aspects of metrology and monitoring of radon, thoron and their daughters.

Phases I and II constitute an integrated set of tasks which have emerged from discussions by the Committee on Radiation Protection and Public Health over recent years. The programme is based on the Proceedings of the NEA Specialist Meetings on Dosimetry and Monitoring of Radon, Thoron and their Daughter Products at Elliot Lake, Canada, in October 1976, Paris/La Crouzille in November, 1978, the NEA International Symposium on "Management, Stabilization and Environmental Impact of Uranium Mill Tailings", Albuquerque, in July 1978 and also on discussion at the Arlington Workshop on Radiation Protection for Naturally Occurring Radionuclides, in May 1978.

The work is intended to provide specialised guidance on dosimetry and monitoring for radiation protection from radon, thoron and their daughters, particularly in uranium or thorium mines, but also in other cases of exposure, both occupational and domestic, for application to and fulfillment of the recommendations of the ICRP.

The publication by the OECD of reports such as this may contribute to the development of an international consensus on matters of public concern. It does not commit Member governments or the Organisation in any way.

## ACKNOWLEDGEMENTS

The contribution of A Birchall (NRPB, Chilton, UK) to the development of dosimetric models for Chapter 2 is gratefully acknowledged and also that of K Eisfeld (GSF, Munich, FRG). Mrs G M Fisher prepared the text using facilities generously provided by the National Radiological Protection Board (UK).

## Members of the Expert Group (Phase I)

| | |
|---|---|
| Australia | Mr. V.A. LEACH |
| Austria | Dr. F. STEINHÄUSLER (Chairman) |
| Belgium | Dr. R. BOURGOIGNIE |
| Canada | Dr. J.R JOHNSON |
| Denmark | Mr. H.L GJØRUP |
| F.R. of Germany | Prof. W. JACOBI |
| Finland | Mr. O. CASTREN |
| France | Mrs. A.M. CHAPUIS<br>Mr. J.F. PINEAU<br>Mr. P. ZETTWOOG |
| Italy | Dr. G. MASTINO<br>Dr. G. SCIOCCHETTI |
| Japan | Mr. M. KINOSHITA |
| Netherlands | Dr. B. BOSNJAKOVIC<br>Dr. L. STRACKEE |
| Norway | Dr. E. STRANDEN |
| Portugal | Dr. M. M. TORRES-SIMOES |
| Sweden | Mr. H. EHDWALL |
| United Kingdom | Dr. A. C. JAMES (Consultant and Editor) |
| United States | Dr. A. GOODWIN |
| CEC | Prof. M. COPPOLA<br>Dr. H. SEGUIN |

## Secretariat

Mr. P.J. RAFFERTY, NEA

The following took part in the second meeting of the Expert Group:

Dr. T. BORAK (United States)

Dr. K. SCHIAGER (United States)

TABLE OF CONTENTS

TABLE OF CONTENTS (cont)

TABLE OF CONTENTS (cont)

TABLE OF CONTENTS (cont)

TABLE OF CONTENTS (cont)

## SUMMARY OF RECOMMENDATIONS

Based on a review and analysis of models for calculating dose absorbed by sensitive cells throughout the bronchial and pulmonary regions of the lung, recommendations of the International Commission on Radiological Protection and current knowledge of environmental variation in the physical characteristics of radon and thoron daughter aerosols, it is concluded that reference values for the conversion factors between exposure to potential $\alpha$-energy and effective dose equivalent can be applied in radiological protection of workers and the general public.

To assess occupational exposure to radon daughters in all underground and open-pit mines, a conversion factor of 8.5 mSv per WLM exposure to potential $\alpha$-energy (2.4 Sv per J h $m^{-3}$) is recommended to represent conditions typical of exposures over an annual period. This conversion factor is also recommended for non-mining occupational exposure to radon daughters, provided that the standard mean breathing rate of exposed workers (1.2 $m^3$ $h^{-1}$) is applicable. At the lower average breathing rate of 0.75 $m^3$ $h^{-1}$, appropriate for adult members of the general public exposed indoors, a reference conversion factor of 5.5 mSv per WLM exposure (1.5 Sv per J h $m^{-3}$) is recommended to assess exposure. This factor can be applied for the whole population without age correction. For both occupational and domestic exposure to radon daughters, dose is absorbed principally by bronchial tissue and an increased risk of inducing bronchogenic cancer is the expected consequence. The probability of inducing bronchogenic cancer per unit effective dose equivalent for exposure in mines is expected to include a component from exposure to co-factors.

For occupational exposure to thoron daughters a conversion factor in the range 2-4 mSv per WLM exposure to potential $\alpha$-energy (0.6-1.2 Sv per J h $m^{-3}$) is recommended. This is about one-third the value recommended for exposure to radon daughter potential $\alpha$-energy. Dose to bronchial epithelium contributes about one-half of this effective dose equivalent, the remainder is contributed by irradiation of bone surfaces and the kidneys. For exposure of the general public to thoron daughters, the factor converting potential $\alpha$-energy exposure to dose is reduced in proportion to the average adult breathing rate. A rounded conversion factor of 0.5 Sv per J h $m^{-3}$ is recommended to assess population exposure to thoron daughters, again without age correction.

The criterion recommended for adequacy of the quantity exposure to potential $\alpha$-energy as a measure of dose in occupational radiological protection is that factors converting exposure to effective dose equivalent assessed for the particular conditions of exposure for an individual worker should not exceed 1.5 x the conversion factors implied by the ICRP in recommending annual limits on exposure of 0.017 J h $m^{-3}$ and 0.05 J h $m^{-3}$ respectively, for radon and thoron daughters. These limits on exposure take into account the contribution of exposure co-factors to the risk of bronchogenic cancer observed amongst underground miners. On this basis, monitoring and control of exposure to potential $\alpha$-energy will provide an adequate means of limiting annual dose if the conversion factor between exposures and effective dose equivalent does not exceed 4.5 Sv per J h $m^{-3}$ for the radon daughters and 1.5 Sv per J h $m^{-3}$ for the thoron daughters. This criterion is generally satisfied in the case of occupational exposure to radon and thoron daughters. It is recommended, therefore, that the determination of individual exposure to potential $\alpha$-energy by appropriate monitoring provides an adequate means of limiting annual dose.

For exposure of the general public to radon and thoron daughters, application of the conversion factors 1.5 Sv per J h $m^{-3}$ and 0.5 Sv per J h $m^{-3}$, respectively, is recommended to set limits on exposure corresponding to limits of risk. In this case, assessment of exposure to potential $\alpha$-energy is again judged to be an adequate basis for limiting population dose.

The objectives of monitoring are to achieve and maintain safe working conditions for occupationally exposed groups and safe environmental conditions for members of the public. Monitoring should therefore enable the effectiveness of environmental control measures to be evaluated. It must also, by means of records of exposure and dose assessment, demonstrate compliance with national dose standards. In the workplace, the balance of effort between environmental sampling and monitoring of individual exposure to potential $\alpha$-energy should be optimised within the local constraints of costs and resources, bearing in mind the levels of exposure likely to be incurred and the overall objective of reducing exposure.

For the purpose of interpreting epidemiological data, unbiased estimates of absorbed dose should be used. These may require investigation of aerosol characteristics in addition to potential $\alpha$-energy in particular exposure situations and also evaluation and recording of any co-factors present in the exposure environment. The scope and justification for such investigations, however, will probably be determined by the precision of the risk estimate obtained from the study population. The basis for assessment of dose in epidemiological studies, as in radiological protection, must therefore be measurement of representative samples to estimate individual exposure to potential $\alpha$-energy.

## Chapter 1

## INTRODUCTION

Epidemiological studies of several populations of uranium miners exposed in the past to relatively high concentrations of radon gas and its daughters during their work underground have demonstrated an excessively high incidence of lung cancer. A proportional relationship between excess lung cancer incidence and exposure to potential $\alpha$-energy from the short-lived daughters of radon is indicated (ICRP, 1981). However, these epidemiological data cannot be used directly to set an occupational limit for inhalation of radon daughters, because the range of risk estimates emerging from the different studies is too wide. This may be due in part to the fact that estimates of exposure of the individual miners constituting each study group are largely retrospective and based on very incomplete or otherwise inadequate air sampling. Exposure to potential $\alpha$-energy (defined in Appendix A) is nevertheless a convenient practical quantity for monitoring and has been widely applied as such and to control individual exposure in uranium mines for more than a decade.

It is generally believed that by improvement of ventilation systems and other management practices and adoption of an annual limit of exposure to radon daughters in the region of 4 Working Level Months (WLM), it is unlikely that the harmful effects of exposure to radon daughters will be observed among uranium miners in the future. Nevetheless, there is considerable justification for the opinion that exposure of uranium miners remains one of the most significant radiological impacts of the nuclear fuel cycle.

In view of the uncertainties in epidemiological data, estimation and control of risks associated with radon daughters in uranium mines requires that exposure should also be evaluated in terms of the basic dosimetric standard of effective dose equivalent (ICRP, 1981). This need is exacerbated by concern in some countries that exposures to radon daughters in ore treatment plants and in non-uranium mines can also be significant, as can exposure of the general public in dwellings constructed of particular materials or on ground exhaling high concentrations of radon gas. Differences in the physical characteristics of aerosols and conditions of exposure between these environments may well be expected to result in different relationships between the quantity 'exposure' and dose.

Exposure to thoron daughters, both occupational and domestic, is also of widespread concern. There are no epidemiological data relating to thoron daughters, therefore assessment of risk must be based entirely on a dosimetric comparison with exposure to potential α-energy from radon daughters (ICRP, 1981).

These problems provided the impetus for Phase I of the NEA programme of work on radon dosimetry and monitoring. The following fundamental requirements were identified at the outset of the work:-

(1) <u>To review dosimetry of radon, thoron and their daughters, and in particular:-</u>

- to examine analytical models for the assessment of dose to individuals who are exposed to aerosol mixtures of these nuclides;

- to examine the state of knowledge of the characteristics of aerosol mixtures of these nuclides and other factors that influence the dose to the lung, and identify those parameters which should be measured for dosimetric purposes in radiological surveillance programmes;

- to examine the variation that might be expected in the factor converting exposure to dose between different conditions of exposure;

- to identify areas where further study or research are needed.

(2) <u>To review the use of exposure as an index of dose, and in particular:-</u>

- to establish criteria for sufficient accuracy in assessment of the relationship between exposure to potential α-energy and dose;

- to establish the limits of physical aerosol parameters or biological factors within which exposure to potential α-energy alone is an adequate correlative of dose, for the purpose of radiological protection.

(3) <u>To review the objectives of measurement and monitoring in specific environments, and in particular:-</u>

- to examine the requirements for individual dose assessment and record keeping for the purpose of radiological protection;

- to examine monitoring requirements for environmental control in workplaces or dwellings;

- to distinguish between requirements for dose assessment in radiological protection and in epidemiological studies.

These three tasks are addressed below in separate Chapters. They were tackled sequentially by three task groups drawn from the membership of the Expert Group.

The review of dosimetric models in Chapter 2, prepared by James, Jacobi and Steinhäusler, is based on the modelling methods described by Jacobi and Eisfeld (1980) together with more recent developments evolved in the course of this work within the framework recommended by the International Commission on Radiological Protection in ICRP Publication 26 (ICRP, 1977). Results of this model review have been taken into account by the ICRP in recommending limits for inhalation of radon and thoron daughters by workers (ICRP, 1981). The United Nations Scientific Committee on the Effects of Atomic Radiation has based its assessment of exposure of the general public on the modelling results given in Chapter 2 for indoor exposure.

The examination of the adequacy of exposure measurement as an index of dose in Chapter 3, prepared by Johnson and Leach, is based on the modelling results of Chapter 2. The emphasis in Chapter 3 is placed on examining limits of applicability of an exposure standard in relation to the basic limitation of dose, for the purpose of radiological protection. Particular reference is made to the exposure limits recommended for inhalation of radon and thoron daughters by workers in ICRP Publication 32 (ICRP, 1981).

Chapter 4, prepared by Steinhäusler, Leach, Stranden and Ehdwall, reviews the range of situations in which exposure to radon and thoron daughters is of concern and monitoring and dose assessment are required. Guidance is offered on any parameters in addition to potential $\alpha$-energy exposure that need to be monitored and recorded in each case and also on the principles that should be observed in designing a cost-effective monitoring programme.

The main conclusions of the report are summarised in Chapter 5.

Chapter 2

# RESPIRATORY TRACT DOSIMETRY OF RADON AND THORON DAUGHTERS

## 2.1 INTRODUCTION

Over the past two decades sophisticated dosimetric models have been developed which enable dose to sensitive tissue in the lung resulting from inhalation of radon and thoron daughters to be evaluated. These models take into account the special physical properties of radon and thoron daughter aerosols which give rise to non-uniform irradiation of lung tissue by $\alpha$-particles.

Until quite recently ICRP models were based on the concept of a critical tissue receiving the highest dose. This concept was adopted by the International Commission on Radiological Protection in their 1959 Recommendations on Permissible Dose for Internal Irradiation (ICRP, 1959). It was introduced to assess dose to lung tissue from exposure to inhaled radon and thoron daughters by Chamberlain and Dyson (1956). These authors calculated that the walls of the trachea and large bronchi would receive the highest dose because of the high deposition efficiency for atoms of polonium-218 (RaA) or lead-212 (ThB) remaining unattached to ambient aerosol particles in inhaled air.

Later calculations employed detailed anatomical models of airway dimensions and branching structure throughout the bronchial tree. These enabled more complete evaluation of the equilibrium concentrations of radon or thoron daughters on bronchial surfaces resulting from the combined effects of deposition of the inhaled aerosol, radioactive decay and transport in the mucus lining the airways. The calculations of Altshuler et al (1964), Harley and Pasternack (1972a) for radon daughters in mine air and Jacobi (1964), Haque and Collinson (1967) for radon daughters in indoor atmospheres thus enabled bronchial dose contributed by the major fraction of the radon daughter atoms attached to ambient aerosol particles that deposit deep in the bronchial tree to be assessed.

These dosimetric models took into account attenuation of $\alpha$-particle energy by the mucous blanket and an insensitive layer of epithelial tissue. Dose to the epithelial basal cells, which were presumed to be radiosensitive, was evaluated at a defined depth below the airway surface. The calculations are well reviewed elsewhere (Walsh, 1970; Walsh and Hamrick, 1977; UNSCEAR, 1977; Jacobi, 1977; Fry, 1977). They agree in the conclusion that basal cells in segmental bronchi, between about the third and fifth bronchial generations, receive the highest dose. However, estimates of this maximum dose per unit of exposure to potential $\alpha$-energy[(1)] vary by more than an order of magnitude. They range from about 0.2 to 12 rad per WLM (2 to 120 mGy per WLM).

A large part of this dispersion in calculated doses arises as a result of the widely different values assumed by these authors for critical parameters of the dosimetric models. The most important of these are the values assumed for the fraction of potential $\alpha$-energy inhaled as unattached atoms and the depths of sensitive basal cells below the bronchial surface. Indeed, when standard values of these parameters are substituted in the models, variation in the calculated maximum doses to basal cells is much reduced (Jacobi, 1977); giving dose conversion factors in the range from about 0.3 to 1 rad/WLM (3 to 10 mGy/WLM). Harley and Pasternack (1982) have recently applied the concept of a critical tissue receiving the highest dose to assess exposure of the general population to radon daughters in domestic environments. They derive for adults a conversion factor of about 0.5 rad/WLM for the dose received by shallow basal cells in the epithelium of segmental bronchi. This approach to dosimetry is being considered by a committee of the United States National Council on Radiation Protection and Measurements (NCRP).

Residual differences in the parameters and methods of calculation employed in the 'critical tissue' dosimetric models have not enabled lung dose to be evaluated on a generally acceptable basis in terms of variations in aerosol characteristics and uncertainties or variability in the biological factors determining lung deposition, clearance and dosimetry. The application of dosimetric modelling to set standards for exposure to radon or thoron daughters (eg, Harley and Pasternack, 1972), or to interpret the radiosensitivity of lung tissue from epidemiological evidence has therefore been of limited value.

A solution to this problem based on dosimetric concepts recommended by ICRP in Publication 26 (ICRP, 1977) has recently been developed by Jacobi and Eisfeld (1980). These authors analysed the sensitivity of calculated lung dose to variation of physical and biological parameters within the framework of a model of dose to the bronchial and pulmonary compartments of the lung which were treated as distinct regions. They introduced a new concept into the dosimetry of bronchial tissue by considering the effect of the variable depth of target cells throughout the bronchial tree. This tends to reduce the dose calculated for the upper airways in comparison with earlier models. It also increases doses calculated in distal bronchi; leading to a more uniform distribution of calculated dose for sensitive cells throughout the bronchial region. Jacobi and

---

(1) Special quantities and units relating to radon and thoron daughter exposure are defined in Appendix A.

Eisfeld utilised the homogeneity in dose to target tissues predicted by their model for both bronchial and alveolar tissue to evaluate the risk of lung irradiation from inhaled radon and thoron daughters in terms of the estimated risk from uniform external irradiation of lung recommended by the ICRP (1977, 1979). They pointed out that calculated doses to target cells averaged over regions of the lung are less sensitive to variations or uncertainties in the biological parameters around which their model is formulated, at least for inhaled radon daughters.

The effect of varying physical and biological parameters on doses calculated throughout the bronchial tree has also been examined by James et al, 1980. These authors, however, formulated a substantially different dosimetric model from that of Jacobi and Eisfeld, on the basis of different interpretations of the available data. It is the objective of this chapter of the report to re-examine the data base needed to model dose to the bronchial and pulmonary regions of the lung and thus to establish reference models to encompass the range of uncertainty or variability in both aerosol and biological parameters influencing the calculation of lung dose from inhaled radon and thoron daughters.

A comparative approach is taken to examine the degree of dispersion in dose estimates that results from choice of dosimetric model and differentiate this from the effects of external factors such as variation of aerosol characteristics between different exposure environments and biological variables within exposed populations. Although somewhat different conclusions would be drawn by applying the 'critical tissue' concept of dose rather than the 'regional lung dose' approach developed here, particularly regarding greater sensitivity to aerosol parameters of the estimates of critical dose, a comparison of these different philosophies is outside the scope of this report.

In this report reference conditions of exposure to radon and thoron daughters in occupational and domestic environments are discussed. Best estimates of the conversion factors between exposure to potential $\alpha$-energy and dose to bronchial and alveolar tissue are proposed in each case and variable or ill-defined parameters introducing significant uncertainty into these relationships are identified. The effective dose equivalent related to radon or thoron daughter exposure in each of the reference environments is derived on the basis of the concepts recommended by ICRP in Publication 32 (ICRP, 1981), by summing the separate risks estimated for irradiation of bronchial and alveolar tissue. Finally, the distribution of dose throughout bronchial tissue and other factors that may influence sensitivity to bronchial irradiation in different environments are discussed.

## 2.2 METABOLIC AND DOSIMETRIC MODELS

### 2.2.1 Geometrical Lung Models

Several detailed studies of the dimensions and branching structure of the human bronchi have been described (Weibel, 1963; Horsfield et al, 1976; Yeh and Schum, 1980). All of these show that the dimensions of airways at a given level of branching vary and that the number of bifurcations to be negotiated by tidal air in reaching the respiratory airways also varies in different parts of the lung. Since it is not practical to model aerosol deposition and clearance for a lifelike bronchial tree, symmetrical approximations of the true branching structure are used. The Weibel 'A' symmetrical lung model has recently been adopted for calculations of radon daughter deposition and clearance (Harley and Pasternack, 1972, 1972a; Jacobi and Eisfeld, 1980; Hofmann, 1982; Wise, 1982). In order to examine the sensitivity of dose conversion factors to airway dimensions, the symmetrical model of Yeh and Schum should also be considered. These authors have developed a 'typical path' representation of the whole bronchial tree in the adult human, based on more complete measurement of a human lung than the Weibel 'A' model. The cast was made within the thorax to minimise dimensional artefacts and preserve the in vivo angles of branching and inclination to gravity of the bronchi. This additional information is useful in calculating the respective contributions to deposition made by impaction and sedimentation.

According to Yeh and Schum, tidal air is distributed to the terminal bronchioles typically by means of 15 bronchial bifurcations. Thus, if the trachea is represented by generation 0, the terminal bronchioles are classified as the 15th bronchial generation.

The bronchial dimensions published by Yeh and Schum correspond with those of an 80 kg man whose lung is inflated to full capacity and therefore need to be scaled to model aerosol deposition in a standard man. In Figure 2.1 the surface areas and volumes of bronchi in the Weibel 'A' model are compared with those of the Yeh-Schum typical path model. The latter has been scaled to give a dead space including the trachea at resting expiratory volume of 169 $cm^3$ (Davies, 1972), instead of the value of 227 $cm^3$ reported by Yeh and Schum for full lung inflation. The corresponding volume for the first 15 generations of the Weibel 'A' lung is 144 $cm^3$. The total surface area of the trachea and 15 bronchial generations in the Weibel 'A' lung is 3950 $cm^2$ compared with 4220 $cm^2$ for the scaled Yeh-Schum model. Both of these values for bronchial surface area are compatible with the 4000 $cm^2$ recommended for Reference Man (ICRP, 1975), although the dead space volumes are larger (cf 110 $cm^3$). Figure 2.1 shows that in the Yeh-Schum model the bronchi between generations 3 and 11, ranging from the segmental bronchi to the bronchioles, are larger than those in the Weibel 'A' lung. This factor strongly influences calculation of bronchial deposition in these two models.

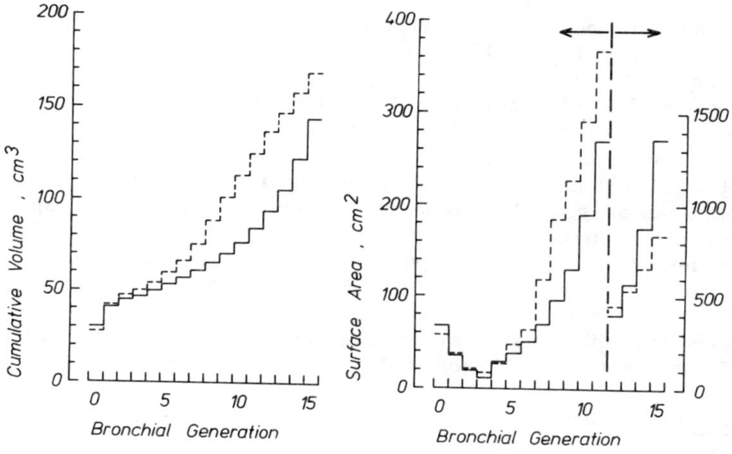


Figure 2.1 Dimensions of model lungs. Cumulative volume and surface areas of the trachea and 15 bronchial generations in the Weibel 'A' lung (——) and a scaled Yeh-Schum typical path model (- - -).

Age dependence

The structure of the bronchial tree is fully developed at birth. Growth occurs by a proportional increase in both airway length and diameter (Hislop et al, 1972). The few data on bronchial dimensions as a function of age (ICRP, 1975) have been reviewed by Hofmann (1982a) in relation to the Weibel 'A' model of the adult lung. Bronchial dimensions in children can be represented by scaling the adult values in proportion to the one-third power of lung capacity,where this increases continuously from about 7.5% of the adult value in a one year old infant to 30% at age six and approximately 50% at age ten (Hofmann, 1982a). The same scaling procedure is applied to derive bronchial dimensions as a function of age from the Yeh-Schum model of adult lung.

## 2.2.2 Breathing rate

Occupational exposure

The tidal volume and respiratory frequency which together determine the breathing rate and intake of potential $\alpha$-energy for a given exposure (Chapter 3 and Appendix A) vary with the physical demands of different work activities. The breathing rate of a miner engaged in heavy work will vary over a range from about 0.45 to 2.4 $m^3\ h^{-1}$ during a daily period of exposure (ICRP, 1975). In the longer term, though, relevant to the assessment of occupational exposure, it is reasonable to characterise intake by a mean breathing rate. Judgement must be exercised in choosing a reference value in the absence of specific occupational data. The mean value of 1.2 $m^3 h^{-1}$ adopted by ICRP (1975) for Reference Man engaged in 'light activity' is probably an over-estimate on the basis of an analysis of dietary intake of energy and oxygen consumption (Adams, N., unpublished). This value seems, however, to be realistic for miners working part of their time at 'high activity' (ICRP, 1981). The value of 1.2 $m^3 h^{-1}$ applies to both underground and open pit miners. For the purpose of calculating aerosol deposition from occupational exposure, it is assumed that a breathing rate of 1.2 $m^3 h^{-1}$ is maintained throughout the exposure period by a tidal volume of 1250 $cm^3$ at a respiratory frequency of 16 $m^{-1}$ (ICRP, 1975).

A somewhat lower mean breathing rate may be appropriate for exposure in factories, mills or radon spas (Chapter 4).

Domestic exposure

A working group of UNSCEAR has proposed on the basis of the ICRP Reference Man Report (ICRP, 1975) the following reference values of mean breathing rates and periods of the day during which adults are exposed indoors and out to radon daughters (UNSCEAR, 1982):

Table 2.1

AVERAGE DAILY BREATHING CONDITIONS FOR ADULT MEMBERS OF THE PUBLIC

| Period | Time h | | Breathing rate $m^3 h^{-1}$ | Intake $m^3$ | |
|---|---|---|---|---|---|
| Indoors | | | | | |
| - Light Activity | 5.5 | 19 | 1.2 | 6.6 | Ca 15 |
| - Intermediate | 5.5 | | 0.75 | 4.1 | |
| - Resting | 8 | | 0.45 | 3.6 | |
| Outdoors | | | | | |
| - Light Activity | 2 | 5 | 1.2 | 2.4 | Ca 5 |
| - Intermediate | 3 | | 0.75 | 2.3 | |
| Mean daily intake ca 19 $m^3$ | | | | | |

It is proposed, therefore, that the rate of intake of potential α-energy in air during indoor exposure is determined by a mean breathing rate of 0.75 $m^3 h^{-1}$, with a higher value of 1 $m^3 h^{-1}$ applying to outdoor exposure. These breathing rates correspond with mean tidal volumes of about 800 $cm^3$ and 1100 $cm^3$ respectively, at a respiratory frequency of 15 $min^{-1}$ (ICRP, 1975).

Age dependence

On the basis of the ICRP Report on Reference Man (ICRP, 1975), the mean breathing rates in 10 year old and 6 year old children, respectively, are represented by 0.6 and 0.5 of the adult values given in Table 2.1 (UNSCEAR, 1982). Similarly, for a 1 year old child an average breathing rate of 0.08 $m^3\ h^{-1}$ (1.4 $l\ min^{-1}$) maintained throughout the day is derived.

Table 2.2

REFERENCE RESPIRATORY PARAMETERS FOR ASSESSMENT OF INDOOR EXPOSURE TO RADON DAUGHTERS

| Age | Tidal Volume $cm^3$ | Respiratory Frequency $min^{-1}$ |
|---|---|---|
| Adult | 830 | 15 |
| 10 year old child | 420 | 18 |
| 6 year old child | 290 | 21 |
| 1 year old infant | 45 | 30 |

Reference respiratory parameters adopted to calculate aerosol deposition in the lung and hence dose from the major component of the population exposure to radon daughter potential $\alpha$-energy incurred indoors are summarised in Table 2.2. Outdoor exposure of the population can generally be neglected in comparison with that indoors.

Table 2.3

ANNUAL EXPOSURE AND INTAKE BY ADULTS PER UNIT CONCENTRATION OF POTENTIAL $\alpha$-ENERGY IN AIR

| Environment | Concentration of potential $\alpha$-energy (1 WL = 2.08 x $10^{-5}$ J $m^{-3}$) | | | |
|---|---|---|---|---|
| | 1 WL | | 1 J $m^{-3}$ | |
| | Exposure WLM | Intake J | Exposure J h $m^{-3}$ | Intake J |
| Occupational | 12 | 0.050 | 2000 | 2400 |
| Domestic | 41 | 0.108 | 6900 | 5200 |
| Outdoors | 11 | 0.038 | 1800 | 1800 |

Annual intake

The annual intake of potential $\alpha$-energy by adults exposed occupationally or as members of the general public in relation to the concentration of potential $\alpha$-energy in air (defined in Appendix A)

averaged over the respective periods of exposure is given in Table 2.3. It is assumed that the exposure of children is the same as that of adult members of the general public, but that intake of potential $\alpha$-energy is reduced pro-rata with the average breathing rate.

### 2.2.3 Deposition in the respiratory tract

In addition to the airway dimensions and the penetration and flow of tidal air in the lung, deposition of radon daughters in the airways is determined by the size distribution of the inhaled aerosol.

#### Unattached atoms

The size and physical behaviour of radon and thoron daughter ions or atoms before attachment to larger airborne condensation nuclei are not well understood (Busigin et al, 1981). Factors such as age since formation, charge, humidity and chemical reaction with gases have led to a range of reported values of the diffusion coefficient for free radon daughter nuclides. A value of about 0.05 $cm^2\ s^{-1}$ is commonly used in deposition calculations, following the early theoretical and experimental work of Chamberlain and Dyson (1956). It is prudent, however, to examine the effect of uncertainty in the diffusion coefficient of unattached daughter species airborne in the respiratory tract on deposition and ultimately dose.

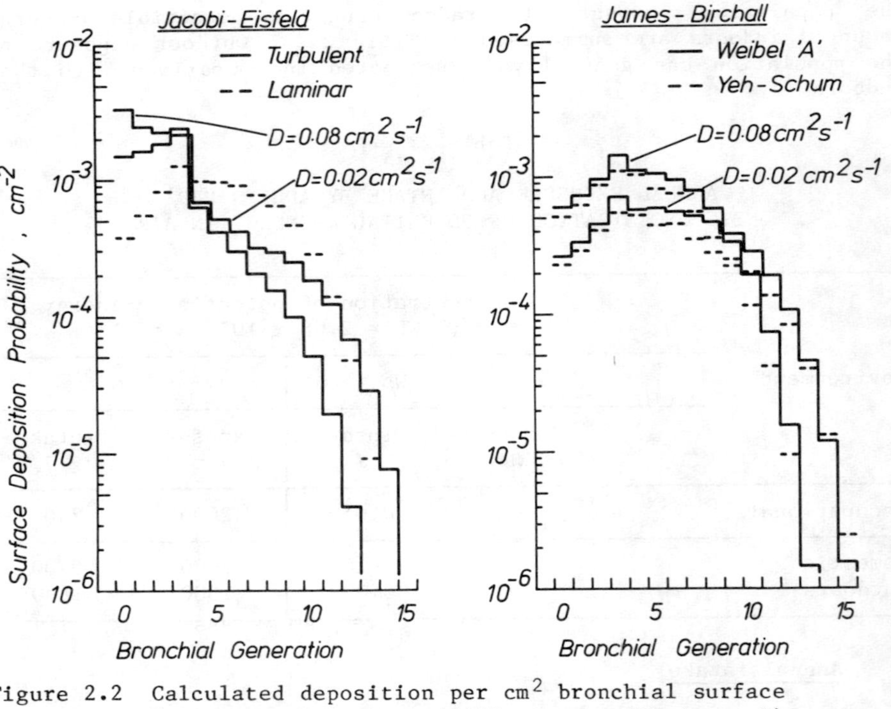


Figure 2.2 Calculated deposition per $cm^2$ bronchial surface for unattached atoms (diffusion coefficient, D) in model lungs. Inhalation through the nose at a breathing rate of $1.2\ m^3\ h^{-1}$.

Figure 2.2 shows the probability of deposition per $cm^2$ bronchial surface calculated for occupational exposure of a miner. The Gormley-Kennedy (1949) formula for diffusion in laminar air-flow is used assuming an extreme range of diffusion coefficient, with the Weibel 'A' and Yeh-Schum lung models. The calculations refer to nose breathing for which nasal deposition is about 50% of the inhaled aerosol (George and Breslin, 1969).

Jacobi and Eisfeld (1980) assumed that deposition in the trachea and first six bronchial generations of the Weibel 'A' lung is higher than the values calculated for laminar flow because of flow disturbance. These authors employed correction factors derived from experimental studies of the deposition of condensation nuclei in a hollow bronchial cast (Martin and Jacobi, 1972). Comparison with the uncorrected deposition values for a median diffusion coefficient of 0.05 $cm^2\ s^{-1}$ in Figure 2.2 shows that the effect of enhancing upper airway deposition is to reduce substantially deposition below the third generation.

In the James-Birchall model (James, A.C. and Birchall, A., unpublished) uncorrected values for laminar flow are used, based on an experimental study of the deposition of unattached $^{212}Pb$ atoms and condensation nuclei 'in vitro' in ventilated pig lung (James, 1977). In these experiments, deposition of unattached $^{212}Pb$ measured in the upper bronchi was less than, but within a factor 2 of, that calculated for an aerosol with diffusion coefficient 0.05 $cm^2\ s^{-1}$. Van der Vooren and Phillips (1982) examined experimentally the deposition in model airway bifurcations of aerosols with median diameters in the sub-micron size range from 0.001 μm (diffusion coefficient 0.05 $cm^2\ s^{-1}$) to 0.2 μm. They concluded that flow disturbance at bifurcations has a negligible effect on deposition in this aerosol size range and also that the equations of Gormley and Kennedy (1949) for diffusion in a straight circular tube are adequate to model deposition in bifurcating airways.

The effect of reducing the particle diffusion coefficient over the range shown in Figure 2.2 is to reduce upper airway deposition, with a compensatory increase in penetration and deposition beyond the segmental bronchi. The larger airways of the Yeh-Schum lung give uniformly lower surface deposition. Penetration of unattached atoms beyond the bronchial airways into alveolar lung is negligible, irrespective of the value of diffusion coefficient or deposition model adopted.

Attached aerosol

The activity median diameter of the radon daughter aerosol formed by attachment to condensation nuclei and dust particles in mine atmospheres, as measured by diffusion batteries, has been reported to vary from about 0.1 μm to 0.3 μm (George et al., 1975; 1977). These authors also reported variable geometric standard deviations, $\sigma_g$, of the aerosol size distributions, with an estimated mean value of 2.7. Bronchial deposition calculated on the basis of different model assumptions for this range of aerosol sizes is illustrated in Figure 2.3, for a mean breathing rate of 1.2 $m^3\ h^{-1}$.

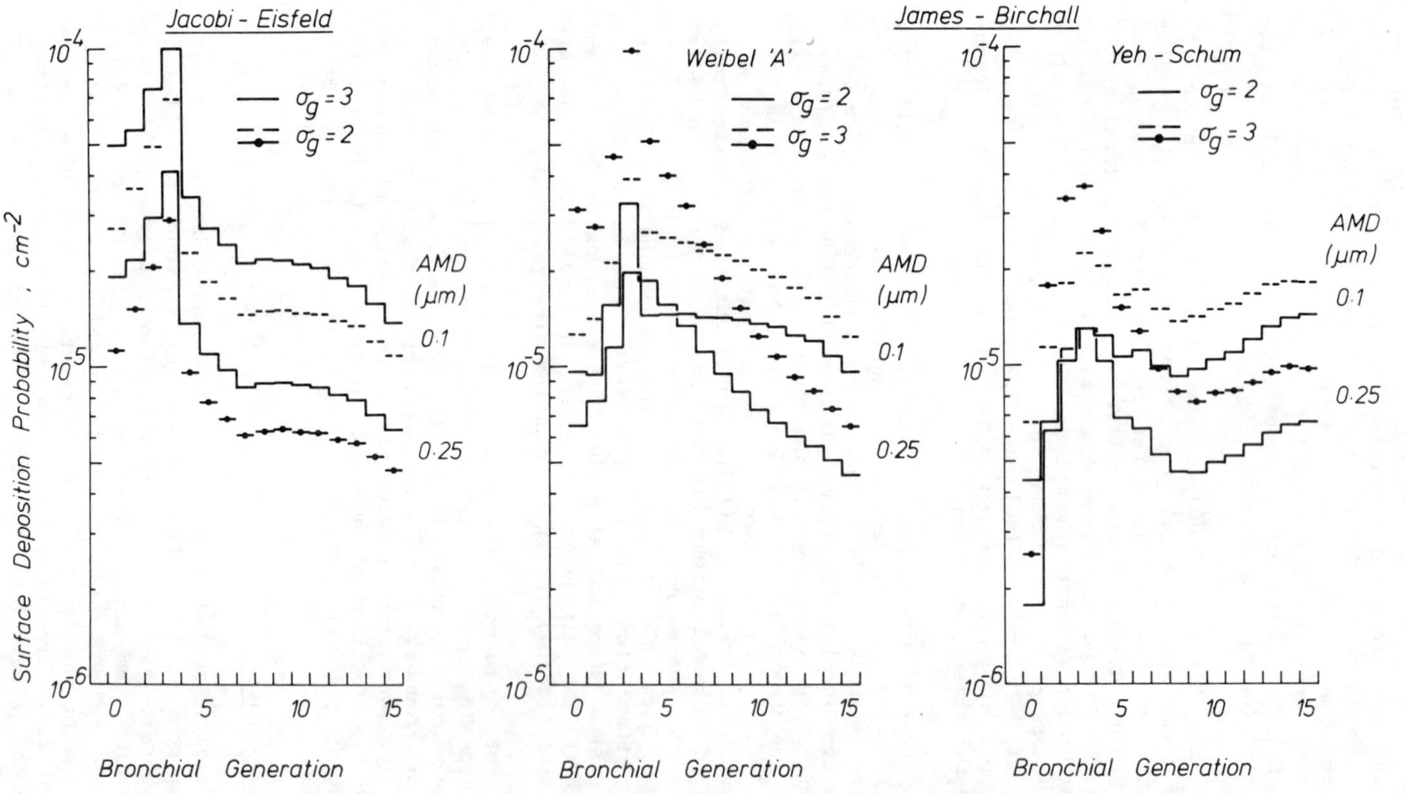


Figure 2.3 Calculated deposition per $cm^2$ bronchial surface for 0.1 μm and 0.25 μm AMD aerosols in model lungs showing the effect of size dispersion, $\sigma_g$. Nose breathing at 1.2 $m^3$ $h^{-1}$.

Jacobi and Eisfeld (1980) assumed a size distribution of 0.25 μm activity median diameter (AMD) with geometric standard deviation, $\sigma_g$, of 3, using the Gormley-Kennedy diffusion equation corrected for higher deposition because of disturbed flow in the upper bronchi of the Weibel 'A' lung. Their analysis is extended in Figure 2.3 over the size range 0.1 to 0.25 μm AMD with $\sigma_g$ of 2 and 3. In this range bronchial deposition by diffusion is uniformly reduced as the aerosol size is increased. The chosen value of $\sigma_g$ is significant, since deposition is up to 50% higher for $\sigma_g$ of 3 than for $\sigma_g$ of 2.

The James-Birchall calculations for the Weibel 'A' and Yeh-Schum lung, also given in Figure 2.3, include the effects of concurrent deposition by impaction (Yeh, 1974) and sedimentation (Heyder, 1977). For a 0.1 μm AMD aerosol, deposition is predominantly due to diffusion. For the larger 0.25 μm AMD aerosol, however, impaction in the upper airways gives a markedly peaked distribution of bronchial deposition. This has little effect on total deposition, which is predominantly bronchiolar and alveolar. Upper airway deposition is substantially higher in the Weibel 'A' lung with its smaller airways and correspondingly higher air velocities than the Yeh-Schum model. Deposition by impaction is rather critically dependent on the assumed value of $\sigma_g$ since, for example, for a 0.25 μm AMD aerosol, as $\sigma_g$ increases from 2 to 3, the fraction of activity carried by particles larger than 1 μm increases from 2.5% to 10%.

Jackson et al, 1982 have studied the association of radon daughter activity with various components of the atmospheric aerosol in underground uranium mines. Using cascade impactors they have distinguished three aerosol components in addition to free atoms (the unattached fraction):

(i) Condensation nuclei with AMAD's < 0.1 μm. Activity attached to this component is significant when the concentrations of larger particulates (eg, dust or diesel soot) are relatively low.

(ii) Diesel soot with AMAD's between 0.2 and 0.4 μm and $\sigma_g$ approximately 2.

(iii) Ore dust with an AMAD of approximately 0.6 μm and $\sigma_g$ of 2.

The activity size distribution of the radon daughter aerosol in mines will depend on the relative magnitude of each of these components.

Davies (1974) has also concluded that a wide dispersion in the size of atmospheric particles arises only from addition of different components with maximal individual $\sigma_g$ values of 2. In the evaluation below of the effect of activity median diameter on dose to bronchial tissue, a value of 2 is assumed for $\sigma_g$ in the James-Birchall calculations, to avoid unrealistically high contributions from inertial deposition of the fraction associated with large particles.

The analysis of Jacobi and Eisfeld (1980) for a 0.25 μm AMD aerosol, assuming $\sigma_g$ of 3 with deposition by diffusion only is extended below over the size range 0.1 to 0.3 μm AMD. It is assumed that $\sigma_g$ of 2 is more appropriate for the small aerosol size distribution.

Effect of breathing rate

Exposure of adults in domestic environments is characterised by a lower mean breathing rate than occupational exposure in mines (Table 2.1). Figure 2.4 illustrates the effect of reducing the breathing rate from 1.2 $m^3\ h^{-1}$ to 0.75 $m^3\ h^{-1}$ on the fractional deposition of unattached atoms per $cm^2$ bronchial area of the Weibel 'A' lung model. With the Jacobi-Eisfeld calculation, deposition in all bronchi beyond the first generation is lower at 0.75 $m^3\ h^{-1}$ than at 1.2 $m^3\ h^{-1}$. In the James-Birchall model, however, without enhancement of deposition in the trachea and major bronchi from the values given by Gormley-Kennedy diffusion, fractional deposition is lower at 0.75 $m^3\ h^{-1}$ breathing rate only in the bronchioles. It is significantly higher in the bronchi.

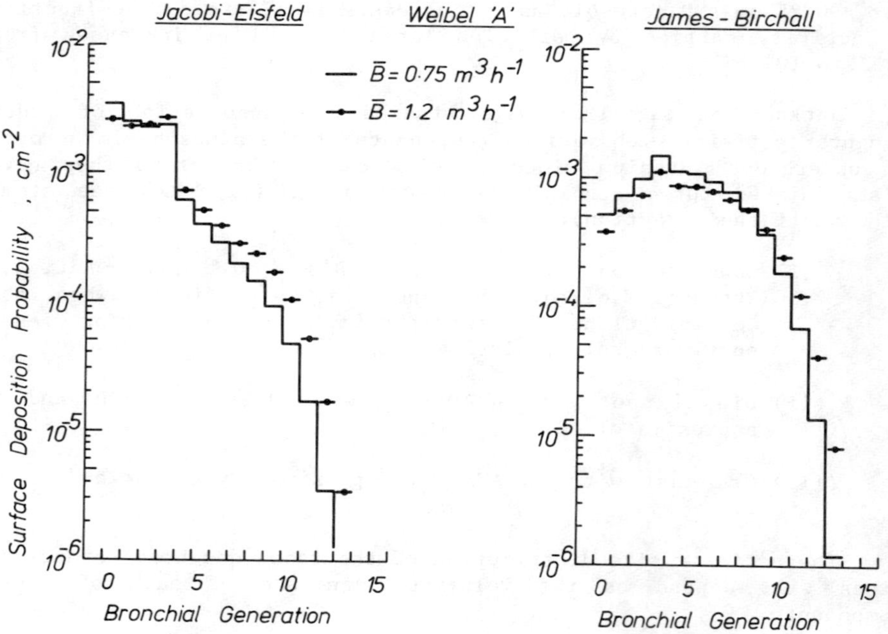


Figure 2.4 Effect of breathing rate on calculated deposition per $cm^2$ bronchial surface for unattached atoms.

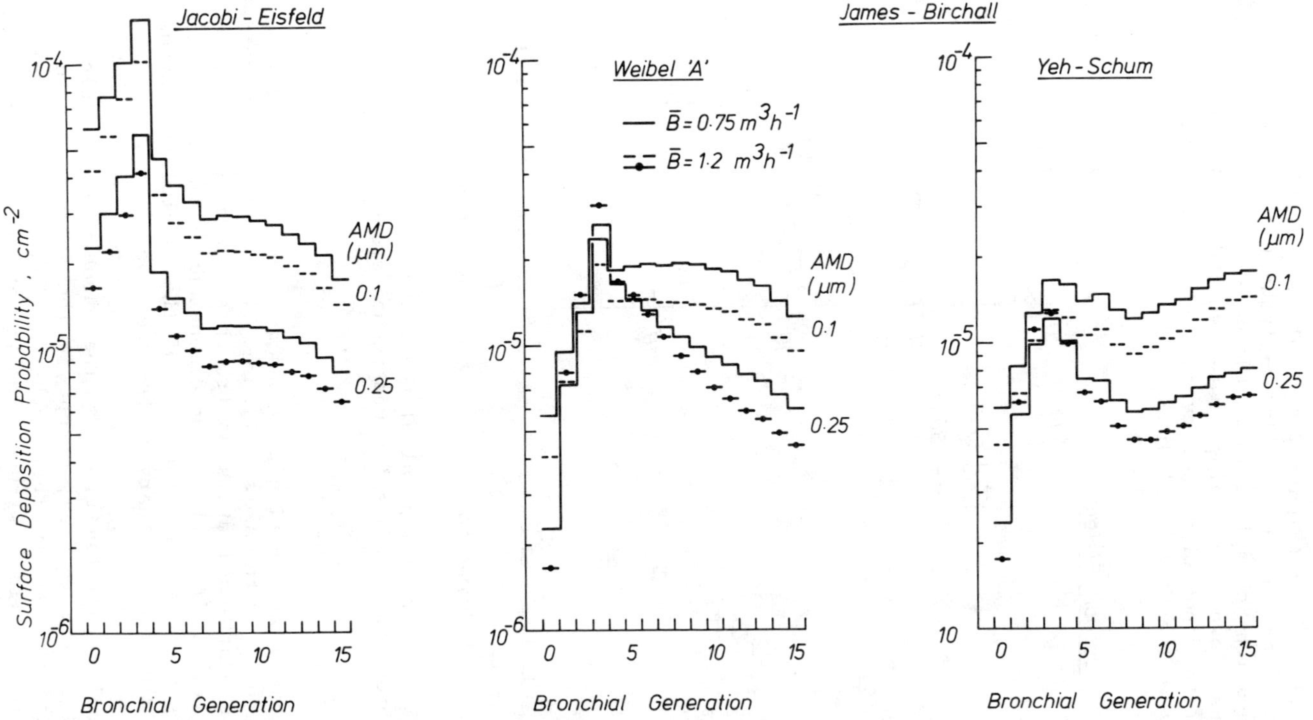


Figure 2.5 Effect of breathing rate on calculated deposition per $cm^2$ bronchial surface for 0.1 μm and 0.25 μm AMD aerosols.

Considering deposition by diffusion only (Jacobi-Eisfeld) fractional deposition in the Weibel 'A' lung is uniformly increased over the size range 0.1 to 0.25 μm AMD as the breathing rate is lowered to 0.75 $m^3 h^{-1}$ (Figure 2.5). The same applies for a 0.1 μm AMD aerosol according to the James-Birchall model for both the Weibel 'A' and Yeh-Schum lungs. In the case of the 0.25 μm aerosol, however, the lower breathing rate substantially reduces the contribution to deposition made by impaction in the upper airways, particularly in the Weibel 'A' lung.

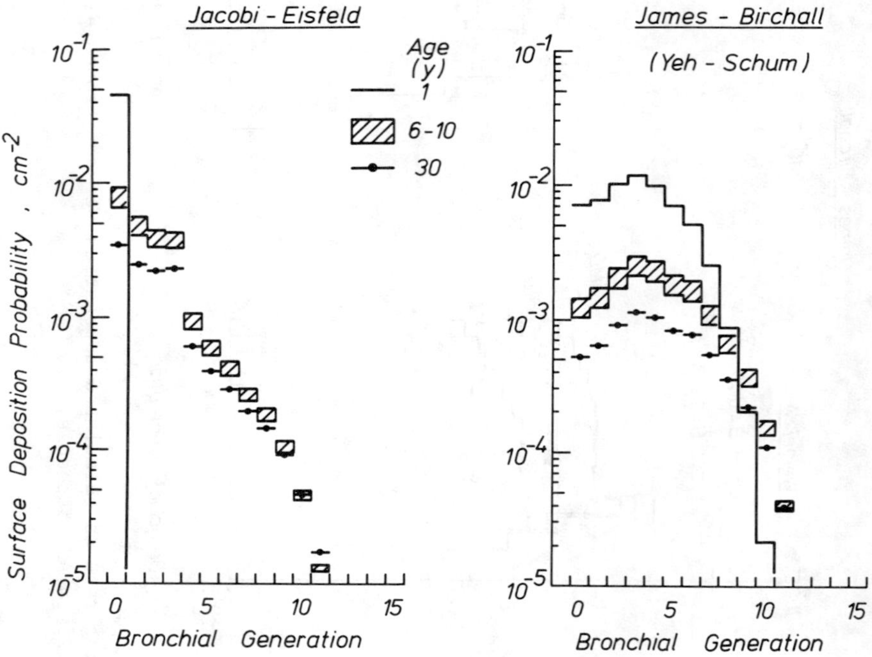


Figure 2.6 Effect of age on calculated deposition per $cm^2$ bronchial surface for unattached atoms.

Age dependence

Both airway size (Section 2.1) and mean breathing rate (Section 2.2) are reduced with age. Figure 2.6 shows the fractional deposition of unattached atoms on bronchial surfaces calculated by scaling the Weibel 'A' adult lung dimensions. In the absence of specific data it is assumed that 50% of the inhaled free atoms are deposited in the nose and pharynx, irrespective of age.

Fractional deposition given by the Jacobi-Eisfeld calculation increases as age is reduced, except in the infant. In this case deposition is complete in the trachea. When the Gormley-Kennedy equations are applied without modification (James-Birchall), fractional deposition in the bronchi increases by a greater amount as the age of the child is reduced, even for the infant.

In Figure 2.7 the fractional deposition of 0.15 μm AMD aerosol, assumed to characterise the attachment of radon daughters in domestic environments (section 5.1 below), calculated by the Jacobi-Eisfeld method as a function of age is compared with that derived for the Yeh-Schum lung (James-Birchall). Both methods give deposition probabilities which increase uniformly by the same relative magnitude as the age of the child is reduced.

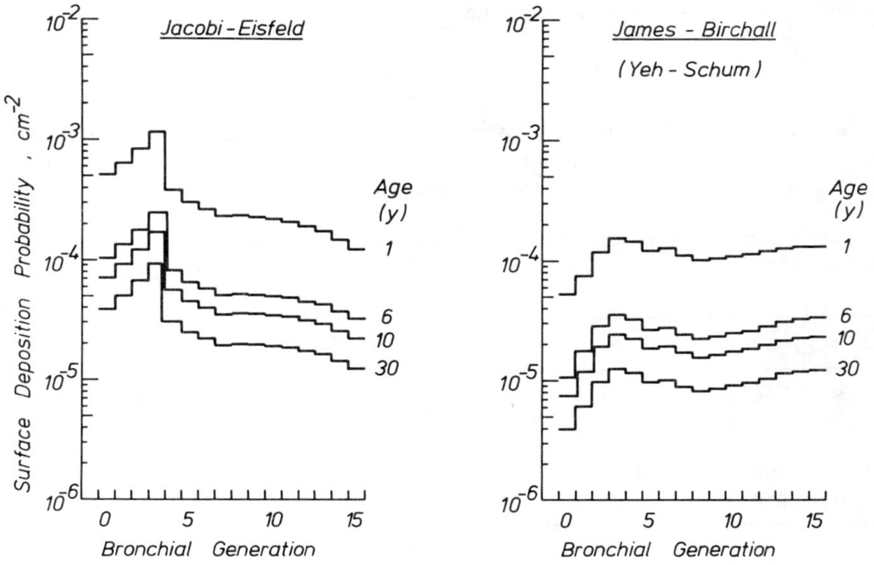


Figure 2.7 Effect of age on calculated deposition per $cm^2$ bronchial surface for 0.15 μm AMD aerosol.

### 2.2.4 Retention in the Respiratory Tract

Retention of radon and thoron daughters deposited in the respiratory tract is determined by the interaction of a number of processes:-

(a) Radioactive decay

(b) Movement through the bronchial region by ciliary transport

(c) Release of attached daughters from their carrier particles and transfer to the systemic blood by diffusion through the epithelium in both the bronchial and alveolar regions

(d) Possible retention of daughter atoms or ions in mucus or mucosal tissue at the site of deposition

Mucociliary transport

Jacobi and Eisfeld (1980) adopted a model in which the velocity of the mucociliary escalator decreases with airway diameter. They estimated a range of mucus velocities in each bronchial generation of the Weibel 'A' lung from reported values of the velocity of single mucus streams in the trachea (1.8 to 1.0 cm $min^{-1}$), judgement of the range of times taken to clear all deposited particles from the human bronchi (0.25 to 2 days) and a power law relationship between mucus velocity and bronchial diameter. Thus:

$$U_i = \frac{l_i}{t_i} = k\, d_i^n \qquad \text{........ (2.1)}$$

where $U_i$, $l_i$, $t_i$ and $d_i$ are velocity of mucus, lengths of bronchi, transit time and diameter of bronchi in generation, i, respectively.

The parameters k and n are determined by the following boundary conditions:

| Ciliary clearance | Fast | Medium | Slow |
|---|---|---|---|
| Mucus velocity, $U_o$ (cm $min^{-1}$) in the trachea | 1.8 | 1.5 | 1.0 |
| Total transit time (d) through T-B region | 0.25 | 0.5 | 2 |

Rate constants, $\lambda_i$, for ciliary clearance in each bronchial generation were derived by Jacobi and Eisfeld (1980) by assuming that the mean-life of material in transit is given by $l_i/U_i$, ie, $\lambda_i = 1/t_i$.

The mean-life of material deposited in a bronchus is assumed to be half that of material entering by ciliary clearance from the distal bronchi, because of the shorter mean transit length.

The range of ciliary transport velocities derived by Jacobi and Eisfeld is shown in Figure 2.8. Also illustrated for comparison are more recent experimental data on the variation of mucociliary clearance between 10 healthy non-smoking subjects (Yeates et al, 1981). In deriving these data a different relationship was assumed between mucus velocity and airway diameter, constrained to keep the thickness of the mucus blanket constant throughout the bronchial tree:

$$U_i = U_o \frac{d_o}{2^i\, d_i} \qquad \text{........ (2.2)}$$

where $U_o$, $U_i$, $d_o$ and $d_i$ are the velocities of mucus and diameters for the trachea and bronchial generation, i, respectively.

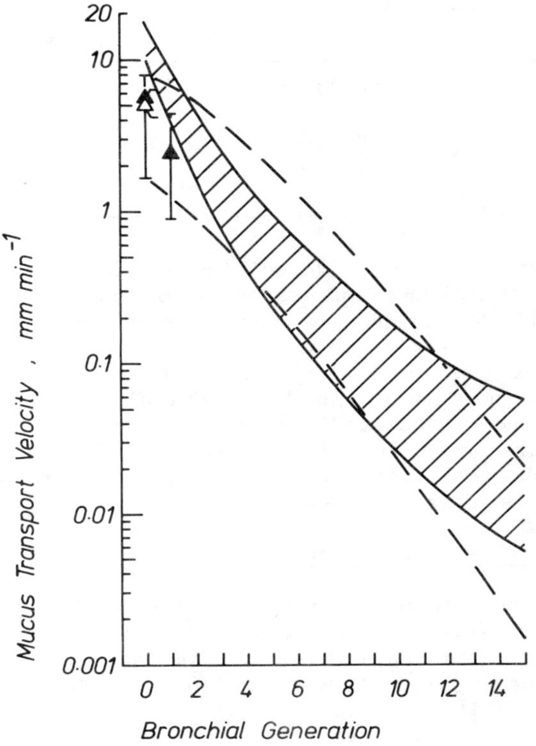


Figure 2.8 Variation of mucus transport velocity with bronchial generation. ▨ Jacobi-Eisfeld model, - - - derived from Yeates et al, 1982, (▲) Foster et al (1981).

The mean transit time through the trachea was measured directly by Yeates et al by observing the passage of boli of activity previously deposited by inhalation of an insoluble aerosol. The mucus velocities throughout the bronchial tree were derived from the tracheal velocity by matching the observed time to clear fractions of the bronchial deposit with the theoretical profile of aerosol deposition. The data are consistent with the mean mucus velocities in the trachea and main bronchi measured by Foster et al (1981), also by means of a non-invasive aerosol exposure.

The inter-subject variation of mucus transport rates throughout the bronchial tree was derived by Yeates et al from the observed variation of tracheal velocity. Corresponding variation in the overall time taken to clear the bronchial tree of deposited particles, derived by applying equation 2.2 to the Weibel 'A' and Yeh-Schum lung models, are as follows:

| Ciliary clearance | Fast | Mean | Slow |
|---|---|---|---|
| Tracheal mucus velocity (mm $min^{-1}$) | 8 | 5 | 1.7 |
| Overall bronchial clearance time (d) | | | |
| Weibel 'A' | 0.7 | 1.0 | 3.1 |
| Yeh-Schum | 0.6 | 0.9 | 2.8 |

Except for generally lower velocities in the trachea and main bronchi, the experimental data of Yeates et al generally confirm the pattern of mucociliary transport throughout the bronchial tree used by Jacobi and Eisfeld. It must be noted, however, that this range of mucus velocities and continuous nature of mucociliary clearance applies to healthy non-smoking adults. Chronic heavy exposure to cigarette smoke and associated bronchitis result in clearance abnormalities before onset of obstructive lung disease (Albert et al, 1971). Such abnormalities include intermittent retrograde mucus flow and intermittent stasis followed by accelerated clearance.

Age dependence of ciliary clearance

Variation of the rates of mucociliary clearance with age is unknown. It is probably most reasonable to assume, however, that the velocity of mucus in the trachea and in each bronchial generation is independent of age (Crawford, 1981), with similar inter-subject variation. This assumption implies that the rate of ciliary clearance for each bronchial generation is higher in children than adults, since the transit lengths are scaled according to the cube root of lung volume.

Compartment models

The processes determining retention of radon daughters act concurrently in each bronchial generation and in the alveoli. To evaluate retention it is convenient to represent these mechanisms by rate constants acting on compartments in competition with radioactive decay.

Jacobi-Eisfeld model

Daughter atoms or ions attached to condensation nuclei, $D_i$, are deposited in the $i^{th}$ bronchial generation in a desorption compartment, $S_i$ (Figure 2.9). The rate at which ions are released from the carrier particles and enter a biological compartment, $B_i$, in the course of absorption into the bloodstream is represented by a rate constant, $\lambda_s$. Whether the deposited attached daughters are carried by mucus into the next bronchial generation, transfer to tissue or decay in situ in mucus is determined by the relative magnitudes of the biological and physical rate constants. Jacobi and Eisfeld assumed long biological half-times of desorption in the range 0.2-1 day, based on their interpretation of the slow absorption into the blood observed in humans after deposition of attached thoron daughter aerosols (Jacobi et al, 1957; Hursh et al, 1969). Thus,

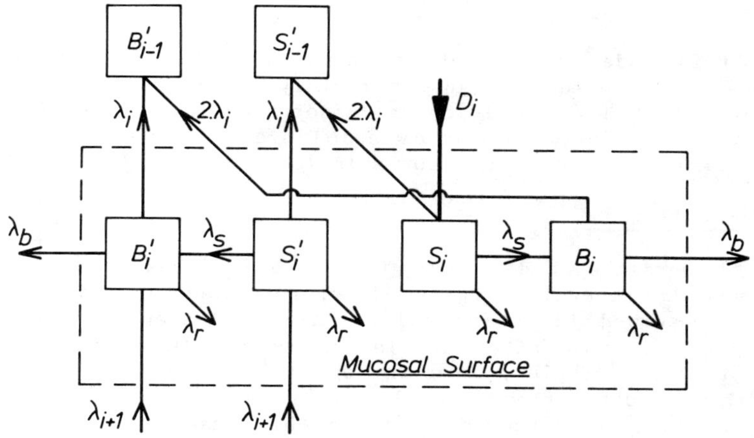


Figure 2.9 Jacobi-Eisfeld model of desorption and transfer in the $i^{th}$ bronchial generation (symbols defined in the text).

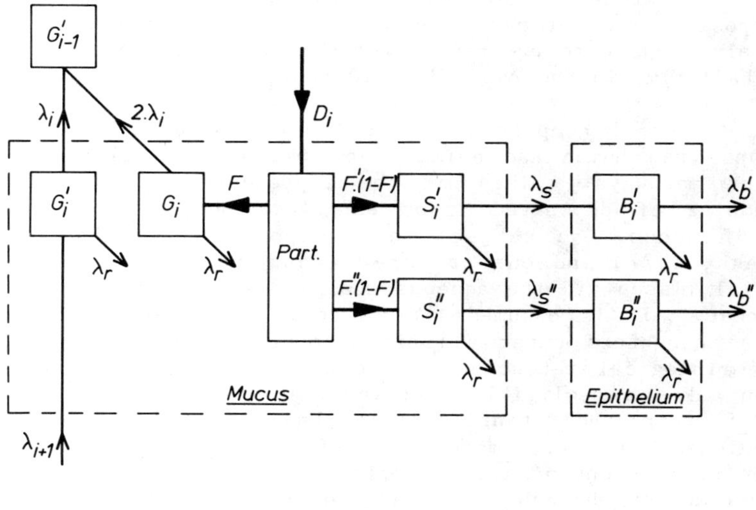


Figure 2.10 James-Birchall model of desorption and transfer in the $i^{th}$ bronchial generation (symbols defined in the text).

for all of the radon and thoron daughters except $^{212}Pb$ (ThB) the value of $\lambda_s$ has little effect on retention at the bronchial surface.

In this model all daughter atoms deposited free of carrier particles are assumed to transfer rapidly to epithelial tissue. A range of half-times for biological absorption $T_b$, between 4 min and 40 min was considered by Jacobi and Eisfeld (1980), based on the ICRP recommendation of 15 min for soluble nuclides (ICRP, 1979).

<u>James Birchall model</u>

The generalised model shown in Figure 2.10 has been formulated to accommodate several different interpretations of radon daughter retention observed in humans and animals. Clearance of a soluble $^{203}Pb\ (NO_3)_4$ aerosol after inhalation by man (Chamberlain et al, 1978) and free $^{212}Pb$(ThB) ions in dogs (Bianco et al, 1974) occurs principally by absorption into the blood, but at a protracted rate similar to that of attached $^{212}Pb$ in the human. The presence of binding sites for radon daughter ions at the bronchial and alveolar surface can therefore be inferred. Some rapid uptake into the blood has also been reported however; for radon daughter ions attached to condensation nuclei in the whole lung of rodents (Pohl, 1962); for lead in sub-micron aerosols of engine exhaust in man (Chamberlain et al, 1978); for free $^{212}Pb$ ions in the bronchi of the rabbit (James et al, 1977); and for $^{212}Pb$ attached to insoluble particles in the nose of the rat (Greenhalgh et al, 1982). Taken as a whole these observations suggest that radon daughter ions may well be desorbed rapidly from carrier particles, but that both the released ions and ions deposited free of particles that follow the absorption pathway are mainly subject to protracted retention at the bronchial surface with a half-time in the range 3 to 10 hours.

The observed complexity in the time course of absorption of lead ions from human and animal lung implies partitioning between different pathways. This may be represented (Figure 2.10) by partitioning of deposited radon daughters between a viscous gel phase of mucus, $G_i$, subject to transport on the tips of cilia (Sturgess, 1977) and one or more transport compartments, $S_i$, to epithelial tissue via the aqueous, sol, phase of mucus in which the cilia beat. It is assumed that a small fraction (about 10%) of deposited daughters enter a rapid absorption pathway with a half-time of biological retention in the mucosa of about 15 min. This fraction makes a negligible contribution to lung dose. Two categories of protracted retention at the bronchial surface may then be distinguished for approximately 30% of deposited daughters; rapid transfer with retention in epithelial tissue (epithelial retention) or slow transfer through the mucous sol (mucous retention) followed by rapid absorption into the blood. It is assumed that a transfer half-time of about 10 hours characterises this protracted retention of radon and thoron daughters in the human lung (Booker et al, 1969; Hursh and Mercer, 1970).

Age dependence of uptake to the blood

It is considered unlikely that uptake of radon daughters by bronchial tissue and subsequent transfer to the blood will be influenced significantly by age.

### 2.2.5 Dosimetric Models

Target tissues

Histological study of lung cancers in US uranium miners has shown that tumours were mainly of bronchogenic origin (Saccomanno, et al, 1981). The histological types of bronchogenic cancer among these uranium miners bear a complex relationship to radiation exposure, cigarette smoking, ageing and induction-latent period. The small cell undifferentiated type predominates in men who started mining early in life, developed respiratory cancer within 20 years and were light cigarette smokers. These tumours were mostly found in the segmental bronchi (Archer, 1978). The epidermoid type was more common in men who started mining after the age of 40 and smoked heavily. However, in cases of unusually high exposure to radon daughters and cigarette smoke, the adenocarcinoma type was most common. The Czechoslovakian studies have also shown that incidence of the small cell undifferentiated and epidermoid types of bronchogenic cancer is related to cumulative exposure to radon daughters (Kunz et al, 1979), but no data are available on the interaction of cigarette smoking.

For the initiation of bronchogenic cancer, bronchial epithelium and particularly the basal stem cells are considered the tissue at greatest risk, although for the purposes of this report dose to alveolar and bronchio-alveolar tissue is assumed to contribute additional risk (ICRP, 1980, 1981).

Depth distribution of basal cells

A large source of uncertainty in dosimetry of radon daughters is the depth of target cells below the bronchial surface, since this is variable and of the same order as the range of $\alpha$-particle emissions (Figure 2.11). Jacobi and Eisfeld (1980) assumed a continuous decrease in the depth of basal cells below the base of cilia and calculated dose at the mean depth for each bronchial generation. The only direct measurements of basal cell depth in the human have been made by Gastineau et al (1972) from a small sample of surgical specimens of 'normal' bronchi. Their data indicate that variability even within five classes of bronchi is very large (Figure 2.11). The mean depths of basal cells in the main, lobar and segmental bronchi may be substantially greater than the Jacobi-Eisfeld assumptions. James and Birchall have used the weighted mean dose to basal cells according to the depth distributions for the five groups of bronchi distinguished by Gastineau et al. According to these authors there are no recognisable basal cells in the bronchioles and therefore stem cells are assumed to occur throughout the depth of the epithelium.

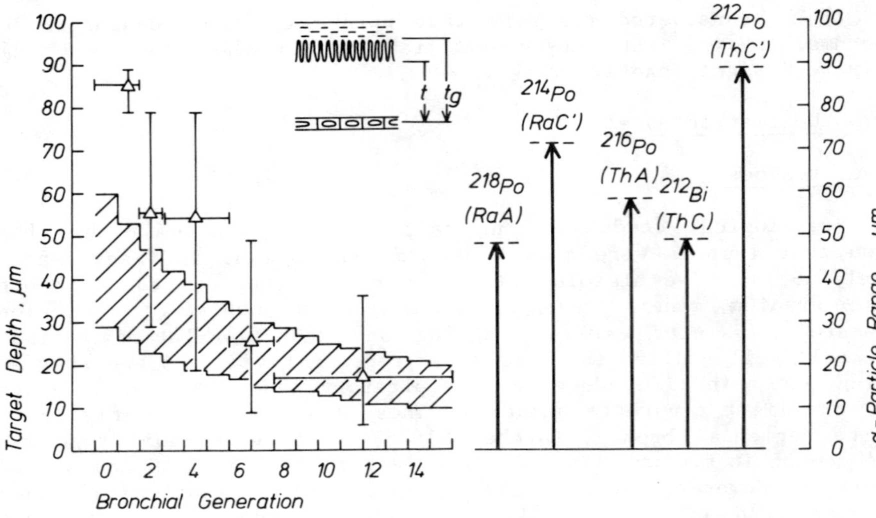


Figure 2.11 Variation in depth of bronchial basal cells, t, beneath 15 μm thick mucus assumed by Jacobi-Eisfeld (hatched) compared with mean depths and variation, $t_g$, beneath 7 μm thick mucus reported by Gastineau et al, 1972 (Λ). Maximum ranges of α-particles emitted in mucus are also shown.

Dose rate to basal cells

Harley and Pasternack (1972) calculated the dose to basal cells from radon and thoron daughters uniformly distributed in a 15 μm blanket of mucus as a function of tissue depth, based on their experimental α-particle stopping powers in tissue equivalent material. The Jacobi-Eisfeld model incorporates these calculated values. James and Birchall have assumed a thinner, 7 μm layer of mucus gel carried on the tips of cilia. Their values, calculated by the Harley-Pasternack method using theoretical range-energy relationships (Armstrong and Chandler, 1973) are shown for comparison in Figure 2.12.

The assumed thickness of active mucus has little effect on dose except at the end of the α-particle range and in the bronchioles where, according to Gastineau et al, the mucus is only 4 μm thick above equally short cilia. The data of Figure 2.12 show that substantially higher doses can be absorbed by stem cells when a concentration gradient of $^{222}$Rn-daughter activity is maintained across the mucosal tissue to the sub-mucosal blood vessels located at about twice the depth of the basal cells. This applies particularly at large depths in tissue.

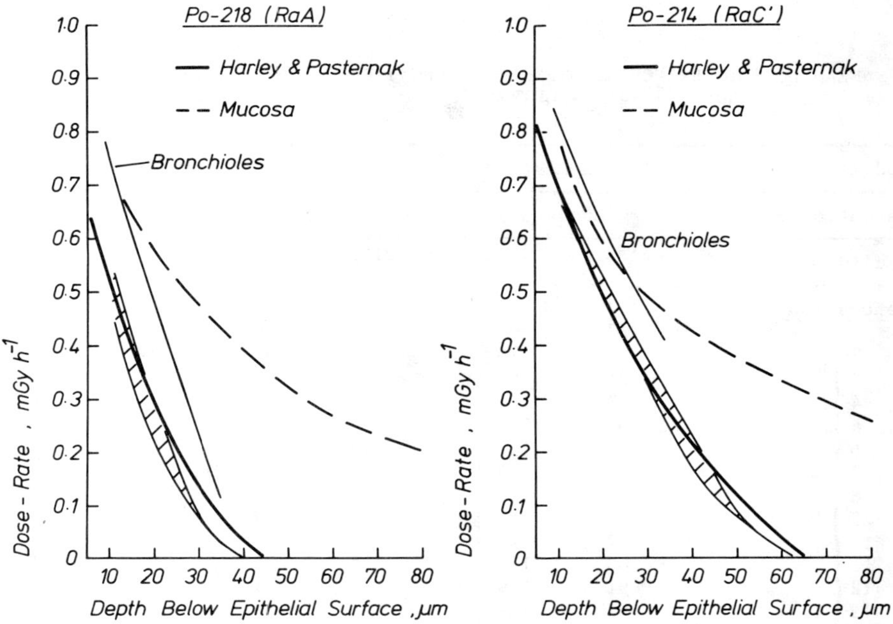


Figure 2.12 Dose-rates in tissue for a surface concentration of 1 Bq $cm^{-2}$ $^{222}Rn$-daughters. Retention in 15 µm thick mucus assumed by Harley and Pasternak (1972). Hatched areas show ranges of dose-rates in bronchial generations 1-7 assuming 7 µm thick mucus on the tips of 7 µm high cilia. Dose-rates for bronchioles assume 4 µm thick mucus and cilia (Gastineau et al, 1972). Dose-rates are also shown for a uniform concentration gradient of activity in mucosal tissue (---).

The dosimetric consequences of these different models are illustrated in Figure 2.13. Mean dose-rates to basal cells in lobar and segmental bronchi from $^{214}Po$ ($RaC^1$) in mucus, weighted according to the distributions of basal cell depths reported by Gastineau et al (the James-Birchall model) are about half the mean values given by Jacobi-Eisfeld. For $^{214}Po$ retained in mucosal tissue, however, dose-rates calculated for the larger bronchi are similar to those of Jacobi-Eisfeld. Differences between the mean doses calculated by Jacobi-Eisfeld and James-Birchall for $^{218}Po$ in the bronchi are greater than those for $^{214}Po$. Dose-rates to bronchiolar stem cells averaged over the whole thickness of the epithelium are approximately a factor two higher than those used by Jacobi and Eisfeld (for both $^{218}Po$ and $^{214}Po$); the site of retention, whether in mucus or mucosal tissue, having no effect.

In Tables 2.4 and 2.5, the mean doses to bronchial stem cells derived by Jacobi-Eisfeld and James-Birchall from their different dosimetric assumptions are compared for the radon and thoron daughters, respectively.

Table 2.4

MEAN DOSE TO BRONCHIAL STEM CELLS (nGy) FROM 1 α-DISINTEGRATION OF THE RADON ($^{222}Rn$) DAUGHTERS PER $cm^2$ BRONCHIAL SURFACE AREA

| Model | Jacobi-Eisfeld | | James-Birchall | | | |
|---|---|---|---|---|---|---|
| Retention Site | Mucus | | Gel | | Mucosa | |
| Nuclide | $^{218}Po$ | $^{214}Po$ | $^{218}Po$ | $^{214}Po$ | $^{218}Po$ | $^{214}Po$ |
| Generation | | | | | | |
| 1 | 17 | 64 | 0 | 0 | 59 | 76 |
| 2 | 27 | 78 | 8 | 41 | 100 | 113 |
| 3 | 35 | 90 | | | | |
| 4 | 43 | 101 | 13 | 50 | 101 | 114 |
| 5 | 51 | 110 | | | | |
| 6 | 58 | 117 | 113 | 161 | 193 | 205 |
| 7 | 64 | 124 | | | | |
| 8 | 70 | 130 | | | | |
| 9 | 76 | 135 | | | | |
| 10 | 82 | 141 | | | | |
| 11 | 87 | 145 | 215 | 229 | 219 | 230 |
| 12 | 91 | 149 | | | | |
| 13 | 97 | 154 | | | | |
| 14 | 101 | 157 | | | | |
| 15 | 106 | 160 | | | | |

Table 2.5

MEAN DOSE TO BRONCHIAL STEM CELLS (nGy) FROM 1 α-DISINTEGRATION OF THE THORON ($^{220}Rn$) DAUGHTERS PER $cm^2$ BRONCHIAL SURFACE AREA

| Model | Jacobi-Eisfeld | | James-Birchall | | | |
|---|---|---|---|---|---|---|
| Retention Site | Mucus | | Gel | | Mucosa | |
| Nuclide | $^{216}Po$ | $^{212}Bi/^{212}Po$ | $^{216}Po$ | $^{212}Bi/^{212}Po$ | $^{216}Po$ | $^{212}Bi/^{212}Po$ |
| Generation | | | | | | |
| 1 | 33 | 72 | 0 | 0 | 67 | 75 |
| 2 | 45 | 84 | 19 | 52 | 110 | 117 |
| 3 | 54 | 95 | | | | |
| 4 | 63 | 103 | 26 | 58 | 107 | 112 |
| 5 | 70 | 111 | | | | |
| 6 | 77 | 119 | 140 | 150 | 196 | 198 |
| 7 | 84 | 126 | | | | |
| 8 | 89 | 131 | | | | |
| 9 | 95 | 137 | | | | |
| 10 | 101 | 142 | | | | |
| 11 | 105 | 145 | 230 | 220 | 260 | 237 |
| 12 | 109 | 149 | | | | |
| 13 | 114 | 154 | | | | |
| 14 | 118 | 157 | | | | |
| 15 | 122 | 159 | | | | |

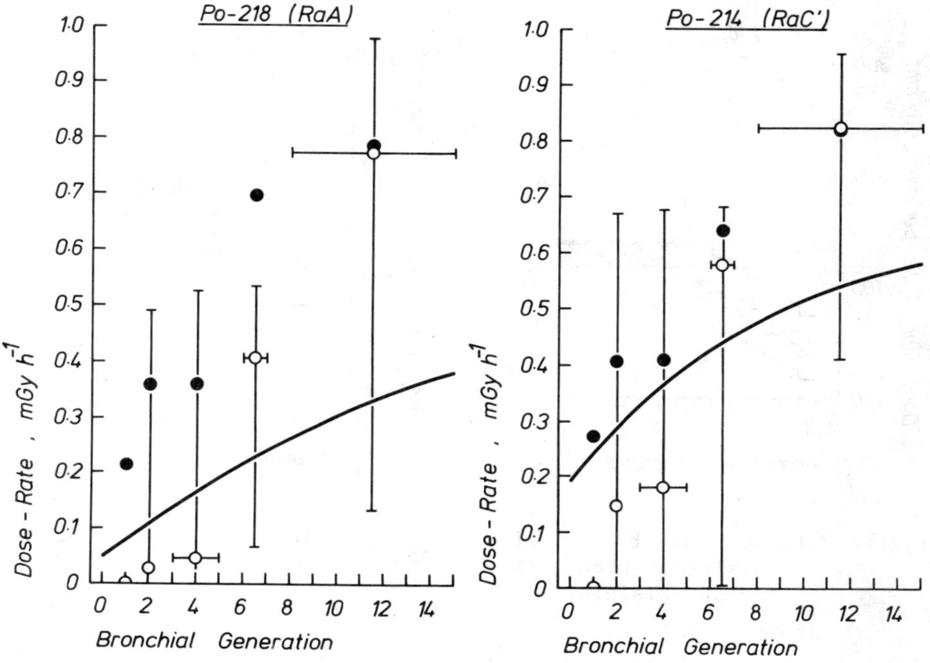


Figure 2.13 Dose-rates to basal/stem cells from 1 Bq $cm^{-2}$ $^{222}$Rn-daughters retained in mucus (Jacobi-Eisfeld, ——; Gastineau et al, ⊢O⊣) and mucosal tissue (●).

Age dependence

The distribution of stem cell depths beneath the layer of mucus as a function of age is unknown, as is the age dependence of mucus thickness. It is probably realistic, however, to assume that basal cell depth decreases with age in proportion to the airway dimensions (Hofmann, 1982) and that mucus thickness, which has a lesser effect on dose, is independent of age. Conversion factors between $^{222}$Rn-daughter activity on the bronchial surface and dose to stem cells as a function of age are illustrated in Figures 2.14 - 2.16.

In order to examine the effect on bronchial dose of this assumed age dependence of target cell depths, doses for children exposed to radon daughters are also calculated below (Section 5.6) using the conversion factors applicable to the adult bronchi (Table 2.4).

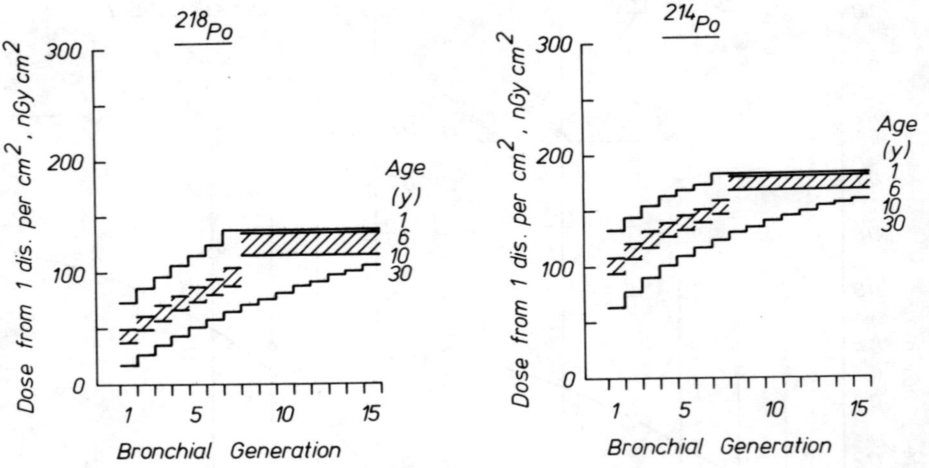


Figure 2.14 Dose to basal cells as a function of age from 1 α-disintegration $cm^{-2}$ $^{222}Rn$-daughters, derived from Jacobi-Eisfeld.

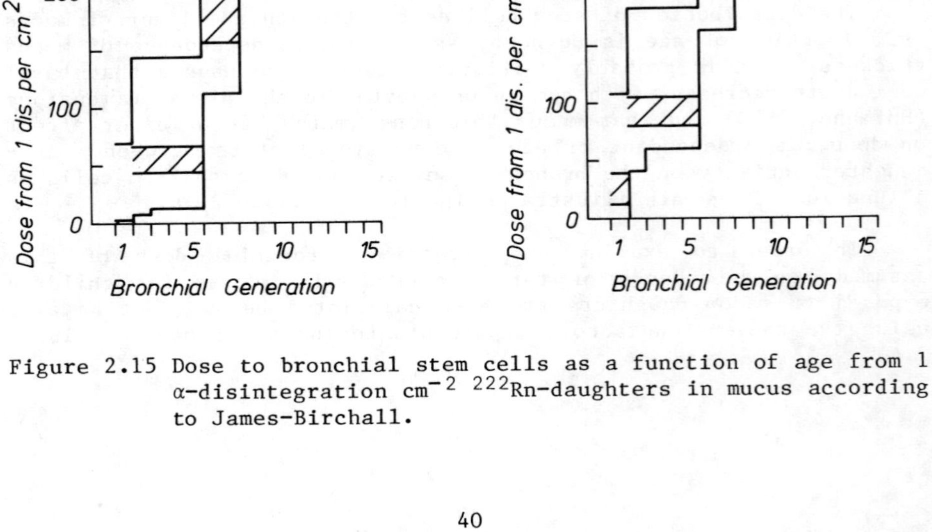


Figure 2.15 Dose to bronchial stem cells as a function of age from 1 α-disintegration $cm^{-2}$ $^{222}Rn$-daughters in mucus according to James-Birchall.

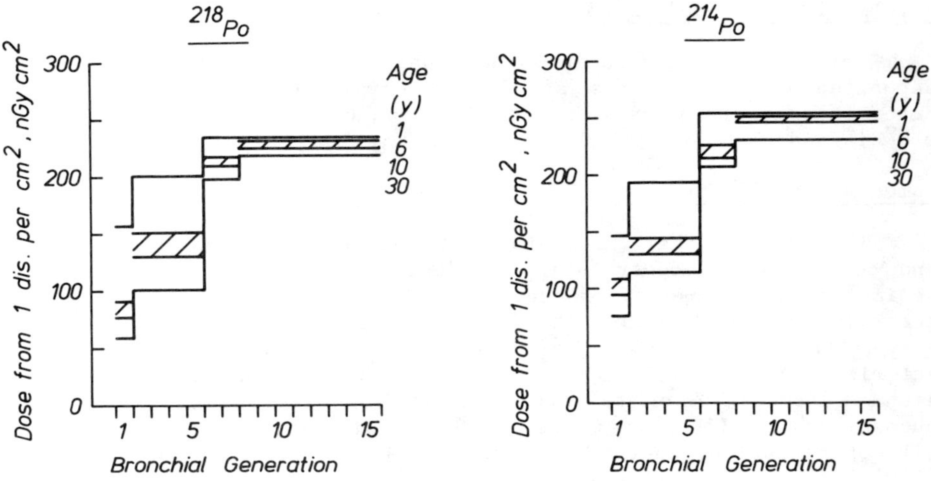


Figure 2.16 Dose to bronchial stem cells as a function of age from 1 α-disintegration $cm^{-2}$ $^{222}Rn$-daughters retained in bronchial tissue according to James-Birchall.

## 2.3 ASSESSMENT OF DOSE AND EFFECTIVE DOSE EQUIVALENT

### 2.3.1 Regional lung dose

The critical cells with respect to induction of lung cancer or other stochastic radiation effects are assumed to be the stem cells of ciliated bronchial epithelium and any other sensitive cells located in alveolar epithelium and the non-ciliated distal bronchioles (ICRP, 1980; 1981). These can be regarded as target tissues in the lung for dosimetry of inhaled radon daughters. The dosimetric quantities related to risk are thus the mean dose or dose equivalent to sensitive cells in the tracheobronchial (T-B) and pulmonary (P) regions of the lung respectively (ICRP, 1981).

A mean dose to bronchial stem cells can be derived by averaging with equal weight doses to basal cells in the lobar bronchi (generation 2) through the stem cells in the fifteenth bronchiolar generation (Jacobi and Eisfeld, 1980; ICRP, 1981). Because of the thin-walled structure of the pulmonary region it is adequate to determine the regional dose by averaging the energy absorbed by a mass of 0.95 kg. Uptake and retention of the short-lived $^{222}Rn$- and $^{220}Rn$-daughters in lymphatic tissue can be neglected.

Regional lung doses are derived in this manner throughout the remaining sections of this chapter. The sensitivity of calculated dose to the selection of biological parameters, represented by the alternative dosimetric models described above, is examined for exposure in different environments. For convenience the dosimetric models are referred to as:

Jacobi-Eisfeld model (J-E)

This employs the Weibel 'A' lung dimensions. Retention of deposited radon daughters at the bronchial surface is principally determined by the rate of desorption from carrier particles (Jacobi and Eisfeld, 1980).

James-Birchall model (J-B)

This uses either the Weibel 'A' or Yeh-Schum lung dimensions and considers the contribution to deposition made by impaction and sedimentation. Dose to bronchial stem cells is calculated strictly according to the stem cell depth distributions given by Gastineau et al (1972). Retention at bronchial surfaces is alternatively characterised by the 'J-E' desorption model or local uptake and retention of about 30% of deposited radon daughters in epithelial and mucosal tissue ('E') or mucus ('M'). Dosimetry appropriate to the site of retention, whether in mucus or tissue, is applied.

### 2.3.2 Effective dose equivalent

The ICRP have recommended that absorbed tissue dose from $\alpha$-particle irradiation, expressed in Gray (Gy), should be multiplied by a Quality Factor 20 to give the equivalent dose of X- or $\gamma$-rays in Sievert (Sv) (ICRP, 1977). The effect of 1 Sv of uniform irradiation of the lung has been assessed by the Commission, from the incidence of lung cancer in Japanese atomic bomb survivors, as a lifetime risk of $2 \times 10^{-3}$. This represents 0.12 of the total risk from 1 Sv whole-body irradiation. Hence 0.12 must be applied to weight the whole-lung dose equivalent to derive an effective dose equivalent (Sv) corresponding with whole-body irradiation.

The ICRP have recommended a procedure to derive an effective dose equivalent when irradiation of the lung is non-uniform, for example following inhalation of $^{222}Rn$- and $^{220}Rn$-daughters (ICRP, 1981; Jacobi and Eisfeld, 1980). As a first approximation, it is assumed that the risk per unit dose equivalent of irradiating bronchial basal cells is equal to that from pulmonary irradiation, ie. a weighting factor of 0.06 is applied to each tissue dose equivalent for summation of risk. This judgement is based on the observation that a large fraction of the tumours in the Japanese atomic bomb survivors arose from bronchial epithelium (Cihak et al, 1974) which received the same dose equivalent as pulmonary tissue.

The contributions to regional lung dose from the free atom and attached fractions of the potential $\alpha$-energy can be combined to derive an effective dose equivalent, $H_E$, per unit intake, J, as follows:-

$$H_E = QF \left[ W_{T-B} \left( f_p \bar{D}^f_{T-B} + (1-f_p) \bar{D}^a_{T-B} \right) + W_P (1-f_p) \bar{D}^a_P \right] \text{Sv J}^{-1}$$

where $QF = 20$, $W_{T-B} = W_P = 0.06$

$\bar{D}^f_{T-B}$ mean basal cell dose from free atoms (Gy $J^{-1}$)

$\bar{D}^a_{T-B}$, $\bar{D}^a_P$ mean regional doses from attached aerosol (Gy $J^{-1}$).

The additional contribution to $H_E$ from irradiation of the lung by $^{222}$Rn gas can be neglected for radioactive equilibrium factors, F > 0.1 (Jacobi and Eisfeld, 1981).

Effective dose equivalents contributed by lung irradiation from inhaled $^{220}$Rn gas and by other organs to which the long-lived thoron daughter, $^{212}$Pb, is translocated have been discussed by Jacobi and Eisfeld (1980). These are summarised in Appendix B.

## 2.4 EXPOSURE TO RADON-222 DAUGHTERS IN MINES

### 2.4.1 Characteristics of Mine Aerosols

#### Unattached fraction

The unattached fraction, $f_A$, of $^{218}$Po (RaA) varies widely in short-term grab samples taken in underground uranium mines, from about 0.004 to 0.4 (George et al, 1975). An approximate, inverse relationship between $f_A$ and the more easily measured airborne concentration of condensation nuclei has been established (Duggan and Howell, 1969; George et al, 1975). Theoretical modelling, however, shows that additional variable factors such as the size distribution of the condensation nuclei, ventilation rate and loss-rate to mine surfaces are also expected to influence the unattached fraction together with the radioactive equilibrium (Jacobi, 1972). It has been shown that a fraction of $^{214}$Pb (RaB) in mine air is unattached to nuclei (Mercer, 1975). This fraction is difficult to measure in the field, but laboratory studies indicate that $f_B$ might typically be about 10% of $f_A$ (Mercer and Stowe, 1977).

#### Attached aerosol

The few data available indicate that the activity median diameter (AMD) of the attached aerosol in operating uranium mines ranges from about 0.09 μm to 0.3 μm (George et al, 1975; 1977 Jackson et al, 1982). George et al noted a correlation between particle size and ambient humidity. An increase in the mean size of nuclei in room air by about a factor 2 as humidity was changed from 20% to about 100% was also observed in the laboratory (Porstendorfer and Mercer, 1978) Instability in particle size of inhaled hygroscopic aerosols is expected to influence deposition in the respiratory tract (Martonen and Patel, 1981).

The natural aerosol in the open air probably has an AMD of about 0.08 μm (Mohnen, 1967).

#### Dependence on mine conditions

In the absence of adequate field data we propose that the range of aerosol characteristics shown in Figure 2.17, based on theoretical considerations (Jacobi, 1972), may be assumed to typify modern conditions in mines. In underground mines three categories of ventilation may be distinguished (high, moderate and low) which, coupled with local variation in the rate of generating atmospheric dust, may lead to the ranges of aerosol size (AMD) shown. Figure 2.17 gives the related variation of the unattached fraction of $^{222}Rn$-daughters expressed as the free atom fraction of the potential $\alpha$-energy concentration, $f_p$. Aerosol conditions in early uranium mines to which the epidemiological data relate are unknown, but expected to lie within the range considered here to represent underground mines.

Short-term variation of both $f_p$ and the aerosol AMD may be greater in open-pit mines, for example, in the vicinity of a high grade ore body. To typify long-term exposure in open-pit mines, however, the limited range of $f_p$ and AMD illustrated in Figure 2.17 is proposed.

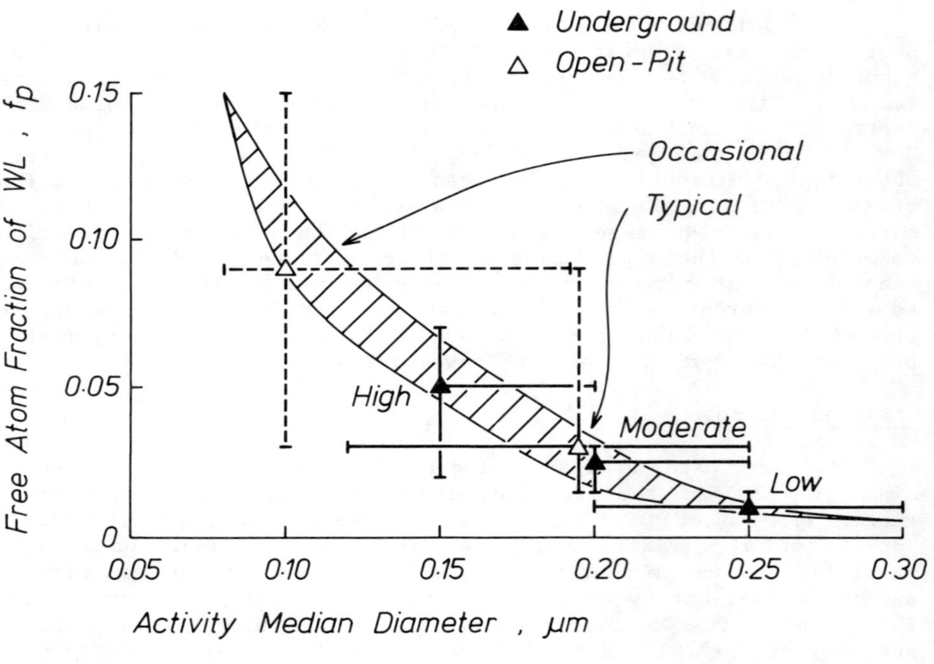


Figure 2.17 Proposed variation with mine conditions of $f_p$ and aerosol AMD for $^{222}Rn$-daughters. Mean values and their typical range are shown for underground mines with high, moderate and low ventilation rates. Temporal variations in $f_p$ and AMD for open-pit mines are also shown. The hatched zone indicates an implied inverse relationship between $f_p$ and AMD.

The radioactive equilibrium factor, F, which represents the ratio of potential α-energy concentration in air to the maximum value if the radon daughters were in radioactive equilibrium with radon gas (Appendix A), may vary in the range from about 0.1 in open-pit mines to 0.7 typical of poorly ventilated underground mines.

### 2.4.2 Regional lung dose

#### Unattached fraction, $f_p$

Figure 2.18 illustrates the range of doses to basal cells, averaged over generations 2-15, calculated for exposure to 1 WLM ($4.2 \times 10^{-3}$J intake) $^{222}$Rn-daughters comprised entirely of unattached $^{218}$Po and $^{214}$Po (ie, $f_p = 1$). The effect of varying the rates of ciliary clearance over the expected human range is shown for each dosimetric model.

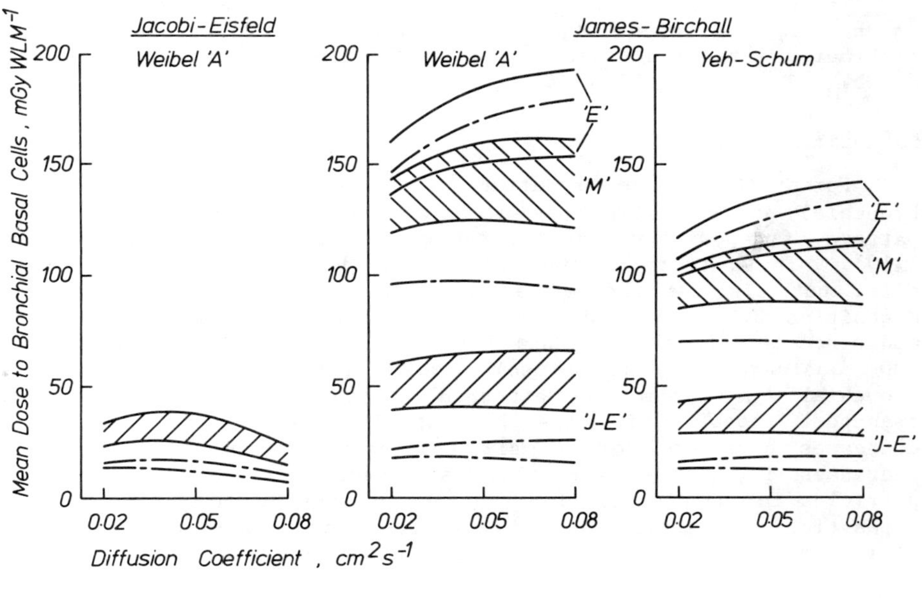


Figure 2.18 Mean dose to bronchial basal cells from exposure to 1 WLM free $^{222}$Rn-daughters; $^{218}$Po (—) and $^{214}$Pb (---). Intake of each nuclide equivalent to $4.2 \times 10^{-3}$ J. Variation of dose with diffusion coefficient, ciliary transport rates and dosimetric model is shown. 'E' refers to 'epithelial retention'; 'M' to 'mucous retention' and 'J-E' to rapid uptake into the blood. At radioactive equilibrium $^{214}$Pb contributes 33% of $f_p$.

The mean dose to bronchial basal cells is insensitive to the assumed diffusion coefficient of the unattached atoms over the extreme range considered. Sensitivity to the rates of ciliary clearance is also minor, all models predicting lower mean bronchial dose for faster mucus transport.

Bronchial dose is very dependent, however, on the absorption characteristics of deposited atoms. The assumption of rapid absorption (biological half-time, $T_b$ = 15 min, 'J-E') leads to about a factor 3 reduction in bronchial doses compared with more protracted retention in either the epithelium or mucus. Transfer and retention in epithelial tissue ('E') gives about 20% higher mean dose than retention entirely in mucus ('M').

The Jacobi-Eisfeld model gives a mean bronchial dose for unattached $^{218}Po$ (RaA) of about 35 mGy $WLM^{-1}$, compared with a range from about 50 mGy to 170 mGy $WLM^{-1}$ for the Weibel 'A' lung and 35 mGy to 120 mGy $WLM^{-1}$ for the Yeh-Schum lung according to the James-Birchall treatment; depending on the assumed retention characteristics.

If rapid absorption is assumed (J-E), doses for unattached $^{214}Pb$ (RaB) are about 50% of the values for $^{218}Po$, whereas more protracted retention gives only slightly lower doses for unattached $^{214}Pb$.

Attached aerosol

For $^{222}Rn$ daughters attached to condensation nuclei, mean bronchial and alveolar dose is more dependent on the calculated pattern of deposition than the assumed biological behaviour (Figure 2.19). The pattern of deposition depends critically on particle size, but also on the method of calculation. Considering only deposition by Brownian diffusion (J-E model), bronchial deposition and hence mean basal cell dose decreases uniformly over the range of AMD considered as the aerosol size distribution becomes larger. Bronchial dose from the attached aerosol increases with mucus transport rates. With the J-B deposition model, bronchial dose decreases over the aerosol size range 0.1-0.3 µm AMD. It then increases for larger aerosols, as impaction of the micron-sized particles in the distribution contributes significantly to bronchial deposition. The influence of deposition by impaction is substantially greater in the Weibel 'A' than the Yeh-Schum lung.

The mean bronchial dose for a 0.1 µm AMD aerosol given by the Jacobi-Eisfeld model is similar to that given by the J-B calculation for the Weibel 'A' lung, since both assumed $\sigma_g$ of 2.

When applied to the J-B dosimetric model, the Jacobi-Eisfeld representation of desorption from particles carried in the mucus gives bronchial doses intermediate between 'epithelial' and 'mucous' retention. In this case the combined effect of variation in ciliary transport and retention characteristics gives only about ± 20% variation in mean bronchial dose.

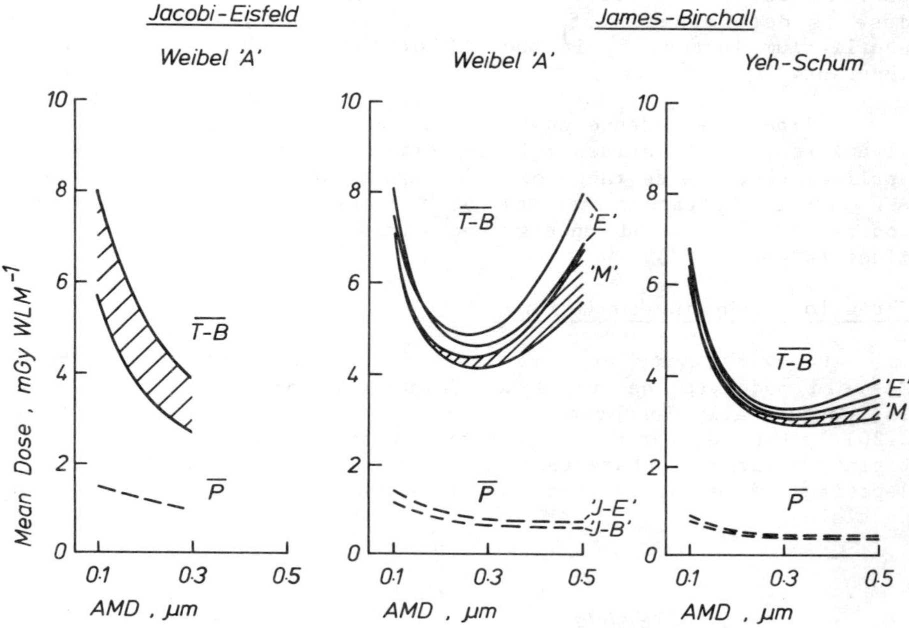


Figure 2.19 Variation of regional lung dose ($\overline{T-B}$, $\overline{P}$) per WLM exposure with AMD of attached $^{222}$Rn-daughter aerosol and dosimetric model, for equilibrium factor $0.05<F<1$. James-Birchall results demonstrate small influence of assumed retention characteristics ('J-E' limited by rate of desorption from carrier particles (not shown) is intermediate between 'E' and 'M').

All of the models give a reduction of about a factor 2 in mean bronchial dose as the AMD of the aerosol is increased over the size-range observed in mine atmospheres, from 0.1 µm to about 0.3 µm AMD. This reported size-range is extended to larger particles in Figure 2.19 in order to identify the trend in lung dose to be expected with possible humid growth of particles in the respiratory tract.

Dose to the pulmonary region again depends principally on the calculated regional deposition and is about 1 mGy WLM$^{-1}$ in the Weibel 'A' model and half this value for the Yeh-Schum lung.

Influence of equilibrium factor, F

Jacobi and Eisfeld (1981) have discussed the state of radio-active equilibrium between the radon daughters and its effect on the conversion factor between exposure to potential α-energy and regional lung dose. Variation of the equilibrium factor, F, is associated with variation in the ventilation rate. Thus, for example, low values of F will be related to high ventilation rates, a higher unattached fraction, $f_p$, and a smaller aerosol size distribution (AMD). The data of Figure 2.20 illustrate that regional lung dose is determined by $f_p$ alone, and is barely influenced by the equilibrium factor, F, if the AMD of the attached aerosol remains constant.

A linear dependence on $f_p$ of the mean dose to the bronchial and alveolar regions arises, irrespective of the dosimetric model applied, over a wide range of F (Figure 2.20). These data assume an AMD for the attached aerosol of 0.2 μm, which may characterise moderately ventilated underground mines and also typify open-pit mines (Figure 2.17).

Variation with mine conditions

The sensitivity of basal cell dose calculated by the Jacobi-Eisfeld model to the unattached fraction of potential α-energy, $f_p$, is substantially less than that of the James-Birchall model (Figure 2.20). This differential sensitivity arises mainly from the different clearance characteristics assumed by Jacobi-Eisfeld for deposited free atoms and radon daughters attached to aerosol particles.

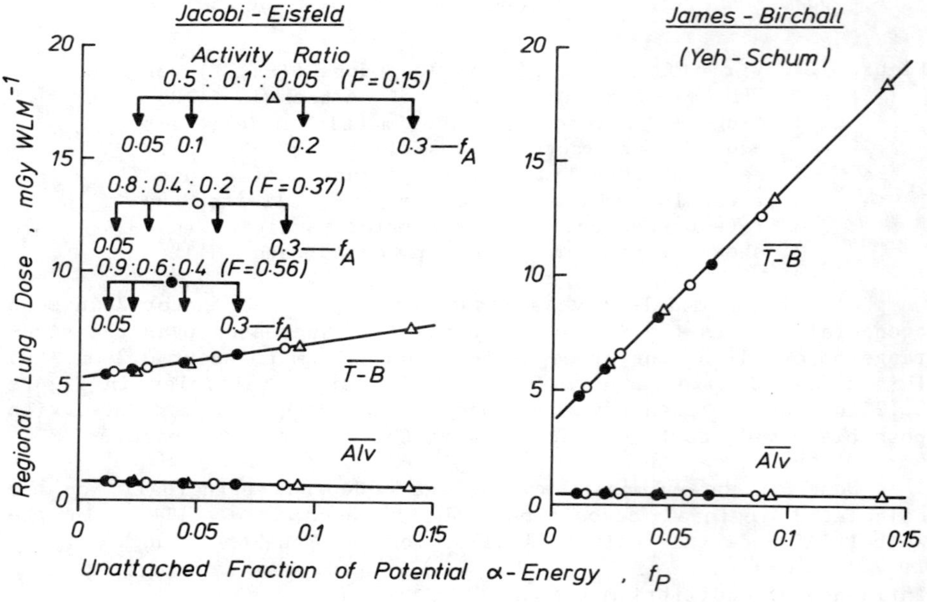


Figure 2.20 Mean dose to bronchial basal cells and the pulmonary region as a function of the unattached fraction, $f_p$, of the inhaled potential α-energy, assuming an aerosol AMD of 0.2 μm, showing that regional lung dose is independent of radioactive equilibrium.

The mean regional doses derived from these dosimetric models as a function of AMD of the attached aerosol (for 1 WLM exposure to potential $\alpha$-energy) are shown in Figure 2.21. Dispersion in these doses for each dosimetric model includes the respective ranges of biological variables. Variability in the dose estimates, however, is due mainly to variation of the physical aerosol characteristics defined in Figure 2.17. The range of AMD and $f_p$ characterising underground mines is extended in Figure 2.21 to include more extreme values of particle size. Thus, smaller aerosols encountered occasionally in open-pit mines are included. The largest aerosol AMD considered (ie, 0.5 μm) may represent the limit of hygroscopic growth in the respiratory tract for the ambient aerosol in a dusty, poorly ventilated mine.

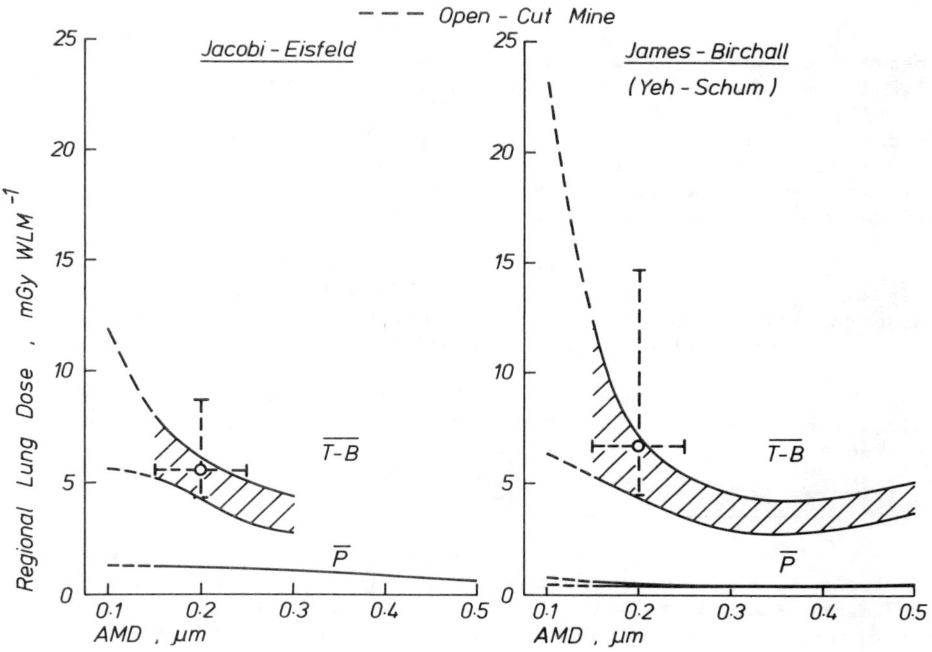


Figure 2.21 Variation of regional lung dose ($\overline{T\text{-}B}$, $\overline{P}$) over the range of $^{222}Rn$-daughter aerosol characteristics encountered in mine atmosphere.

According to the James-Birchall model, variation of the unattached fraction of potential $\alpha$-energy, $f_p$, in mine atmospheres has a larger influence on bronchial dose than the AMD of the attached aerosol, even when possible hygroscopic growth of particles in the respiratory tract is taken into account.

The conversion coefficients between regional lung dose and exposure to potential $\alpha$-energy (WLM, at a mean breathing rate of 1.2 $m^3 h^{-1}$, and also $J h m^{-3}$) are summarised in Table 2.6 in terms of the principal variable, $f_p$. Values given for the James-Birchall dosimetric model relate to the Yeh-Schum lung which probably gives a more realistic assessment of lung-deposition than the Weibel 'A' model. The different sensitivities to $f_p$ of the J-B and J-E dosimetric models have a small practical effect, since the range of $f_p$ in mine atmospheres is very small.

The mean bronchial and pulmonary doses estimated for long-term exposure in underground mines with different ventilation characteristics are compared in Table 2.7 with those estimated for open-pit mines. Ranges of regional lung dose given in the table indicate local variation due to the rate of generating mine dust.

In the case of poorly ventilated mines conversion factors between exposure and mean bronchial dose of about 4 mGy per WLM or 1.2 Gy per $J h m^{-3}$ are derived from both dosimetric models. The coefficients are about 5-6 mGy $WLM^{-1}$ (1.4 - 1.8 Gy per $J h m^{-3}$) in moderately ventilated mines. Higher coefficients of about 7 -10 mGy $WLM^{-1}$ (2.2 - 3 Gy per $J h m^{-3}$) are derived for well ventilated mines; the James-Birchall model giving the highest values because of the greater contribution from free atoms.

Estimates of the mean pulmonary dose range from about 0.5 - 1.2 mGy $WLM^{-1}$ (0.14 - 0.35 Gy per $J h m^{-3}$) with little dependence on mine conditions; the James-Birchall calculation giving values about one half those derived from the Jacobi-Eisfeld model.

Dose conversion coefficients for long-term exposure in open-pit mines are comparable with those for underground mines with moderate - high ventilation.

### 2.4.3 Effective dose equivalent

Variation of the effective dose equivalent, $H_E$, with the particle size distribution typical of mine aerosols is illustrated in Figure 2.22, where both the aerosol AMD and the related free atom fraction, $f_p$, are varied over the ranges given in Figure 2.17. For the James-Birchall model the extreme range of dose conversion coefficients is shown, including maximal values given by the Weibel 'A' lung with epithelial retention of deposited daughters and minimal values given by the Yeh-Schum model with rapid absorption of deposited free atoms.

Figure 2.22 shows that as the aerosol size increases in underground mine atmospheres, the dose conversion coefficients predicted by the two models converge, largely because of the smaller excursions in free atom fraction, $f_p$. The range of $H_E$ that might be expected in an open-pit mine is greater, because of temporal variations in $f_p$. Functions derived from the two dosimetric models relating the effective dose equivalent to potential $\alpha$-energy exposure in terms of aerosol AMD and free atom fraction of potential $\alpha$-energy, $f_p$, are given in Table 2.6. The resulting variation of $H_E$

Table 2.6

CONVERSION FUNCTIONS FOR EVALUATION OF THE REGIONAL LUNG DOSE, D AND EFFECTIVE DOSE EQUIVALENT, $H_E$ PER WLM EXPOSURE FOR WORKERS (NOSE BREATHING AT A MEAN RATE OF 1.2 $m^3$ $h^{-1}$) AND PER J h $m^{-3}$ (1,2,3)

| Conversion Function (1) | Dosimetric Model (2) | AMD = 0.1 μm Lung Region | | AMD = 0.2 μm Lung Region | | AMD = 0.3 μm Lung Region | |
|---|---|---|---|---|---|---|---|
| | | T-B | P | T-B | P | T-B | P |
| D(mGy) per WLM | J-E | 7.1+ 34$f_p$ | 1.5 (1-$f_p$) | 4.6 + 35fp | 1.2 (1-$f_p$) | 3.0 + 36$f_p$ | 1.0 (1-$f_p$) |
| | J-B | 6.5+100$f_p$ | 0.8 (1-$f_p$) | 3.6 +103$f_p$ | 0.5 (1-$f_p$) | 3.0 +104$f_p$ | 0.4 (1-$f_p$) |
| D(Gy) per J h $m^{-3}$ | J-E | 2.0+9.6$f_p$ | 0.43(1-$f_p$) | 1.3 + 10$f_p$ | 0.34(1-$f_p$) | 0.85+ 10$f_p$ | 0.29(1-$f_p$) |
| | J-B | 1.8+ 29$f_p$ | 0.23(1-$f_p$) | 1.0 + 30$f_p$ | 0.14(1-$f_p$) | 0.85+ 30$f_p$ | 0.12(1-$f_p$) |
| $H_E$(mSv) per WLM | J-E | 11+ 39$f_p$ | | 7.1+ 40$f_p$ | | 4.8+ 26$f_p$ | |
| | J-B | 8.8+119$f_p$ | | 4.9+123$f_p$ | | 4.1+124$f_p$ | |
| $H_E$ (Sv) per J h $m^{-3}$ | J-E | 3.0+ 11$f_p$ | | 2.0+ 11$f_p$ | | 1.3+ 11$f_p$ | |
| | J-B | 2.5+ 34$f_p$ | | 1.4+ 35$f_p$ | | 1.2+ 36$f_p$ | |

(1) $f_p$ = unattached fraction of total potential α-energy of the daughter mixture.

(2) J-E = Dosimetric model of JACOBI-EISFELD with Weibel 'A' lung
J-B = Dosimetric model of JAMES-BIRCHALL with Yeh-Schum lung

(3) $H_E$ assumes quality factor, QF = 20 for α-particles and weighting factor, $W_T$ = 0.06 for each of the two target tissues (T-B = bronchial basal cells; P = pulmonary region)

Table 2.7

ESTIMATED LONG-TERM MEAN VALUES OF DOSE CONVERSION COEFFICIENTS FOR MINERS UNDER VARIOUS CONDITIONS(1,2)

| Type of mine and Ventilation Conditions | AMD (μm) | $f_p$ | Dosimetric Model | D(mGy) per WLM | | D(Gy) per J h $m^{-3}$ | | $H_E$ (mSv) per WLM | $H_E$ (Sv) per J h $m^{-3}$ |
|---|---|---|---|---|---|---|---|---|---|
| | | | | T-B | P | T-B | P | | |
| Underground | | | | | | | | | |
| Low ventilation | 0.25 | 0.01 (0.005-0.015) | J-E | 4.2 (4.0-4.3) | 1.1 | 1.2 (1.1-1.2) | 0.31 | 6.4 (6.1-6.5) | 1.8 (1.7-1.8) |
| | | | J-B | 4.2 (3.7-4.8) | 0.5 | 1.2 (1.1-1.3) | 0.14 | 5.6 (5.0-6.1) | 1.6 (1.4-1.8) |
| Moderate ventilation | 0.20 | 0.025 (0.015-0.03) | J-E | 5.5 (5.1-5.7) | 1.2 | 1.6 (1.4-1.7) | 0.32 | 8.0 (7.6-8.3) | 2.3 (2.2-2.4) |
| | | | J-B | 6.2 (5.1-6.7) | 0.5 | 1.8 (1.4-1.9) | 0.14 | 8.0 (6.7-8.6) | 2.3 (1.9-2.4) |
| High ventilation | 0.15 | 0.05 (0.02-0.07) | J-E | 7.5 (6.4-8.2) | 1.2 | 2.2 (1.7-2.4) | 0.35 | 10 (9.1-11) | 2.9 (2.6-3.1) |
| | | | J-B | 9.9 (6.8-12) | 0.6 | 2.9 (1.9-3.5) | 0.17 | 13 (8.8-15) | 3.7 (2.5-4.3) |
| Open-Pit | 0.2 | 0.03 (0.015-0.04) | J-E | 5.7 (5.1-6.0) | 1.2 | 1.6 (1.4-1.7) | 0.32 | 8.3 (7.6-8.6) | 2.4 (2.2-2.4) |
| | | | J-B | 6.7 (5.1-7.7) | 0.5 | 1.9 (1.4-2.2) | 0.14 | 8.6 (6.7-9.8) | 2.4 (1.9-2.8) |

(1) J-E = Jacobi-Eisfeld dosimetric model; J-B = James-Birchall dosimetric model

(2) $H_E$ assumes a weighting factor, $W_T$ = 0.06 for dose equivalent to each lung region

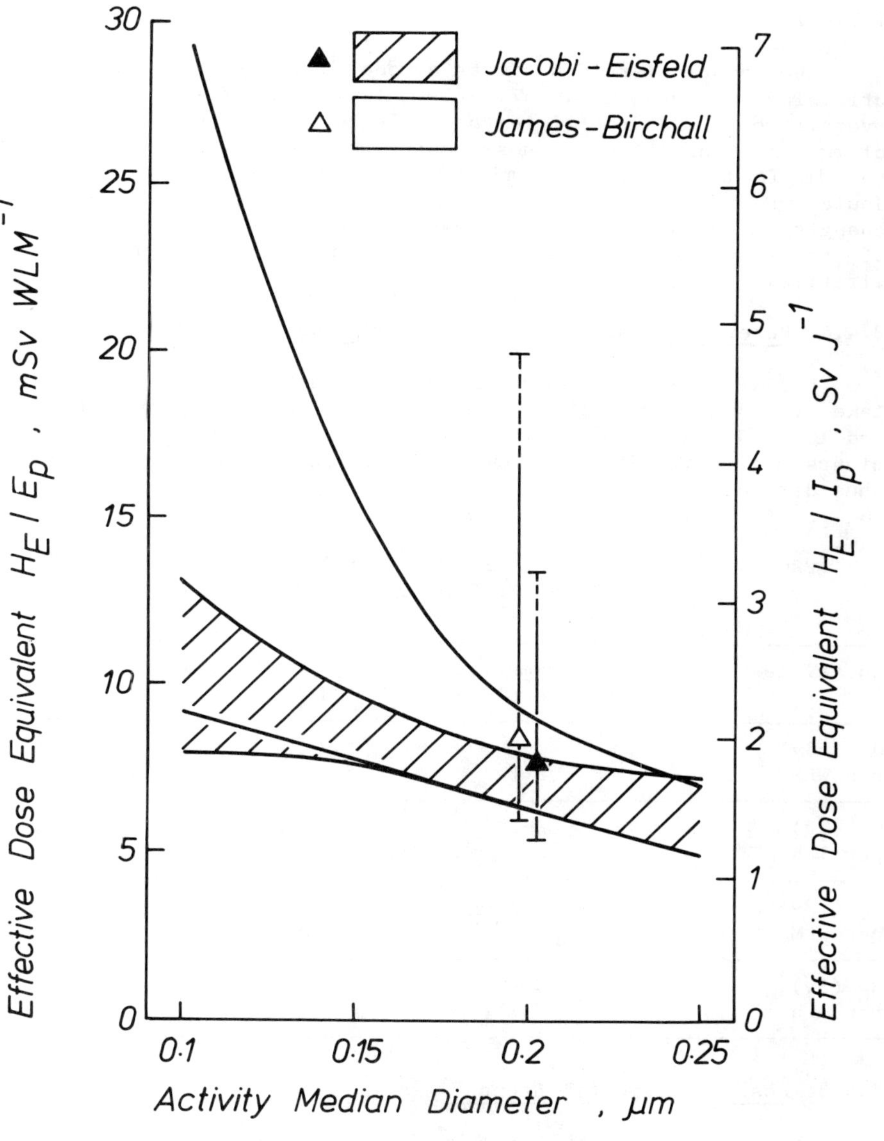


Figure 2.22 Variation of effective dose equivalent, $H_E$ over the range of $^{222}Rn$-daughter aerosol characteristics expected in mine atmospheres. Results for extreme combinations of lung geometry and retention characteristics are shown using the James-Birchall model. The symbols and their error bars illustrate the large short-term variation in $H_E$ expected in open-pit mines.

with conditions in underground and open-pit mines is summarised in Table 2.7

The Jacobi-Eisfeld dosimetric model and that of James-Birchall, utilising the preferred Yeh-Schum lung dimensions, give similar overall conversion coefficients (Table 2.7). For long-term exposure of miners, the effective dose equivalent, $H_E$, is therefore expected to lie in the range 6-10 mSv per WLM exposure (1.5 - 2.5 Sv per Joule intake). Mobility of workers in the mining industry with changing mine conditions is expected to ensure that extreme values of the conversion coefficient will not apply over a working lifetime.

### 2.4.4 Reference dose conversion coefficients for miners

For the purpose of radiological protection it is reasonable to take the average values of the coefficients for regional lung dose and effective dose equivalent given in Table 2.7 as applicable to miners over a working lifetime. The following reference values are thus derived:

Table 2.8

PROPOSED REFERENCE VALUES OF DOSE CONVERSION COEFFICIENTS FOR MINERS EXPOSED TO $^{222}$Rn-DAUGHTERS

| Coefficient | Dosimetric Model | Lung Region T-B | | P | |
|---|---|---|---|---|---|
| $D_\alpha$ (mGy) per WLM | J-E<br>J-B | 5.7<br>6.8 | 6.3 | 1.2<br>0.5 | 0.8 |
| $D_\alpha$ (Gy) per J h $m^{-3}$ | J-E<br>J-B | 1.7<br>1.9 | 1.8 | 0.32<br>0.14 | 0.24 |
| $H_E$ (mSv) per WLM | J-E<br>J-B | 8.2<br>8.8 | 8.5 | | |
| $H_E$ (Sv) per J h $m^{-3}$ | J-E<br>J-B | 2.4<br>2.5 | 2.4 | | |

### 2.4.5 Dose to Segmental Bronchi

It is not known if stem cells throughout the T-B region are equally radiosensitive, for example the number of cells at risk in different bronchial generations and their proliferation rate may vary. The calculated dose to basal cells in segmental bronchi and its dependence on dosimetric model and aerosol size distribution is examined in Figure 2.23 for comparison with the mean regional dose.

Considerable dispersion in the predicted doses arises when the James-Birchall model is used, depending principally on the assumed retention characteristics. Whether or not deposited $^{222}$Rn-daughters are transferred to epithelial tissue is important in this case because most of the basal cells are assumed to be out of range of α-particles emitted in mucus. If it is assumed that free atoms are absorbed rapidly into the blood (J-E), substantially lower doses are estimated for small aerosols typical of clean atmospheres.

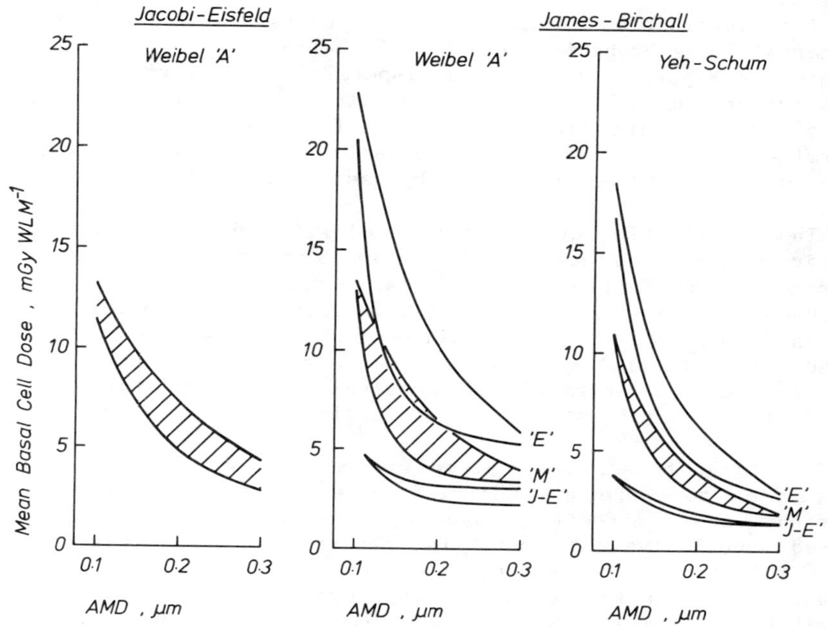


Figure 2.23 Dose to basal cells in segmental bronchi from exposure to 1 WLM $^{222}$Rn-daughters in mines ($4.2 \times 10^{-3}$ J intake). James-Birchall results show influence of assumed biological retention ('E', 'M' or 'J-E'). The reduced variable range for $f_p$ and AMD given in Figure 2.17 is assumed.

If it is assumed that about 30% of deposited daughters are retained in epithelial tissue (J-B model 'E'), basal cells in segmental bronchi are calculated to absorb a similar dose to the regional average . The Jacobi-Eisfeld model also gives this result.

### 2.4.6 Non-parametric factors influencing bronchial sensitivity

Although the effects of additional exposure factors such as smoking cannot be modelled quantitatively, these will probably have a small direct influence on regional lung dose, unless gross changes in the thickness of mucus or the depth distribution of basal cells are caused. Smoking and concurrent exposure to other irritants may well affect the radiosensitivity of bronchial tissue, however. Cigarette tar and other micron-sized chemical carcinogens that may be present in mine air are expected to deposit at relatively high surface concentration in the segmental bronchi (Schlesinger and Lippmann, 1978; James, 1980; Hofmann, 1982). It is likely that this factor has contributed to the absolute risks of radon daughter exposure estimated from epidemiological studies of underground miners (ICRP, 1981).

The distribution of dose to target cells averaged over different parts of the bronchial tree for the Yeh-Schum lung (James, 1982) is shown in Figure 2.24(a). The dispersion of dose estimates shown takes account of the expected variability of both aerosol and biological characteristics. The shape of the dose distribution is determined to a large extent by the variation of target cell depths, which is assumed to be that given by Gastineau et al (1972).

The distribution of absorbed dose, weighted as it is to bronchiolar tissue, differs significantly from the tumour incidence reported by Archer (1978), which is predominantly associated with the shaded region of figure 2.24(a). A characteristically different pattern of deposition pertains for aerosols in the size range associated with potential co-factors in mine air (figure 2.24(b)). In this case, the concentration of deposited mass on bronchial surfaces is calculated to peak strongly by impaction in the upper airways.

If it can be assumed that radiosensitivity of bronchial tissue in relation to cancer is modified by exposure to co-factors, the distribution of effective dose shown in Figure 2.24(c) can be derived speculatively by weighting absorbed dose, in this example, by the relative concentrations of a 1 μm AMAD aerosol of a co-factor deposited on bronchial surfaces. This hypothetical 'effective dose' distribution is more compatible with the observed cancer incidence than the primary absorbed dose.

## 2.5 EXPOSURE OF THE PUBLIC TO RADON-222 DAUGHTERS

### 2.5.1 Characteristics of indoor atmospheres

The physical characteristics of $^{222}$Rn-daughter aerosols in rooms are influenced by the concentration and size distribution of ambient particles and also by the exchange rate between indoor and outdoor air.

#### Room ventilation

Three broad categories of room ventilation in houses can be considered. Typical ventilation rates, representing average conditions over long periods of exposure, may be as follows:

| Ventilation Class | Mean air exchange rate |
|---|---|
| Low | $< 0.3\ h^{-1}$ |
| Moderate | $0.3 - 1\ h^{-1}$ |
| High | $> 1\ h^{-1}$ |

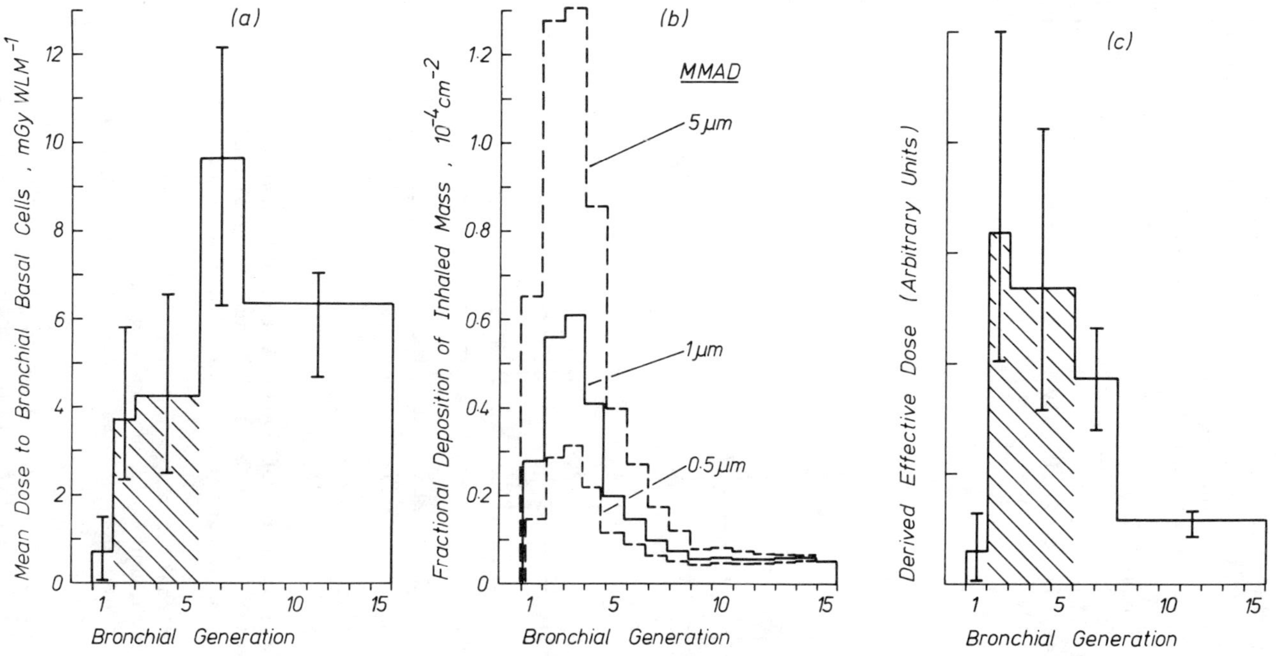


Figure 2.24 (a) Distribution of dose absorbed by bronchial basal cells for exposure in moderately ventilated underground mines (Yeh-Schum lung - basal cell depths according to Gastineau et al).

(b) Pattern of bronchial deposition for chemical agents in mine air as a function of aerosol MMAD.

(c) Distribution of 'effective bronchial dose' derived by normalising the absorbed dose distribution of (a) by the distribution of deposited mass for a 1 μm MMAD chemical aerosol shown in (b).

Aerosol characteristics

There are no reliable experimental data relating to domestic environments. Consideration of the data available, however, supplemented by theoretical modelling, indicates the variation of unattached fraction, $f_p$, and AMD of the attached aerosol with general room ventilation illustrated in Figure 2.25 (Jacobi, 1981). In this figure the mean values of $f_p$ and AMD and their typical ranges for the three classes of house ventilation are compared with values representing conditions in underground mines (Figures 2.17). As in the case of mines, it should be noted that the parameter values for houses may be typical only of long periods of exposure. Temporal variation will undoubtedly be larger.

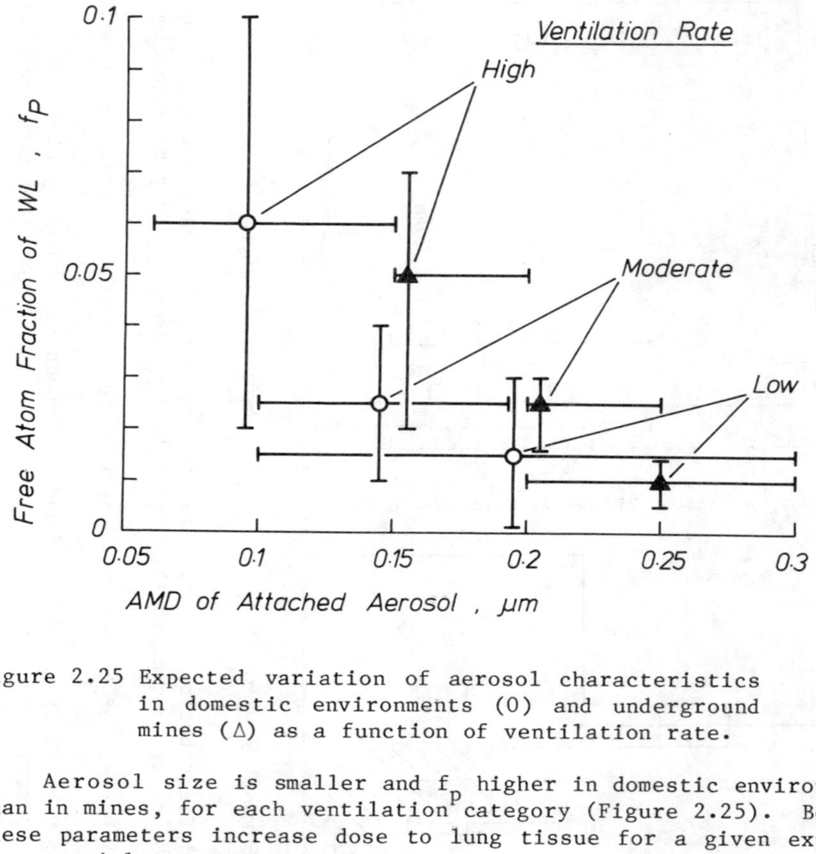


Figure 2.25 Expected variation of aerosol characteristics in domestic environments (O) and underground mines (Δ) as a function of ventilation rate.

Aerosol size is smaller and $f_p$ higher in domestic environments than in mines, for each ventilation category (Figure 2.25). Both of these parameters increase dose to lung tissue for a given exposure to potential α-energy.

The radioactive equilibrium factor, F, varies with room ventilation, but over a smaller range than in mine air. This variable, however, does not influence the conversion from potential α-energy exposure to dose per se (Figure 2.20).

### 2.5.2 Outdoor exposure

The $^{222}Rn$-daughter aerosol in outdoor air may be characterised (Jacobi, 1981) by an average unattached fraction, $f_p$, of about 0.05 (0.02 - 0.08) with an aerosol AMD of about 0.1 μm (0.05 - 0.2 μm). These parameter values are similar to the conditions assumed to typify indoor exposure at high ventilation rates (Figures 2.25). The same dose conversion coefficients will therefore apply.

Lung dose from exposure to potential α-energy out of doors, however, can be neglected for practical purposes in comparison with the much higher exposures incurred indoors.

### 2.5.3 Regional lung dose

#### Unattached fraction, $f_p$

The mean doses to bronchial stem cells in the adult from exposure to the free atom fraction of potential α-energy over the range of breathing rates applicable to indoor exposure, according to various dosimetric models, are shown in Figure 2.26. These data assume that an equilibrium factor, F, of 0.4 is typical of indoor exposure (Keller et al, 1982; UNSCEAR, 1982); but according to the models of regional lung dose the value of F has a small effect on dose. The ranges of dose given in the figure include variability in mucus clearance rates and, in the case of the James-Birchall results, alternative assumptions about biological retention.

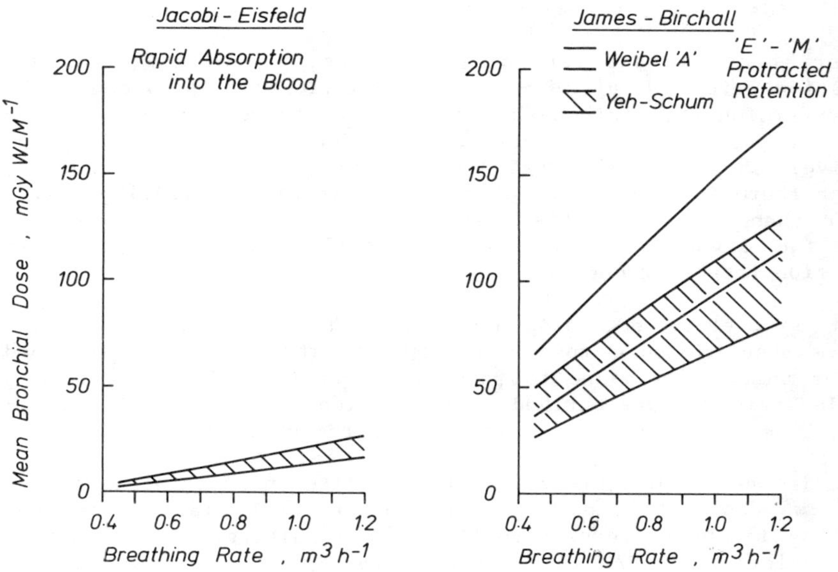


Figure 2.26 Mean dose to basal cells from 1 WLM exposure to free $^{222}Rn$-daughter atoms ($f_p$ = 1) as a function of breathing rate.

Basal cell dose from exposure to free atoms is low according to the Jacobi-Eisfeld model because of the assumed rapid absorption of deposited atoms into the blood. The mean bronchial dose from free atoms varies approximately pro-rata with intake, ie, it is proportional to breathing rate for constant exposure.

Attached aerosol

Figure 2.27 shows the range of doses to the bronchial and pulmonary regions calculated for exposure to 1 WLM potential $\alpha$-energy attached to aerosols. In this case the mean bronchial dose varies less than pro-rata with breathing rate; although alveolar dose shows greater variation. These observations follow from the effects of flow-rate and tidal volume, respectively, on deposition. The influence of variable and uncertain biological parameters on bronchial dose from the attached aerosol is small.

Variation with room ventilation

Regional lung doses for exposure to 1 WLM potential $\alpha$-energy, derived as a function of breathing rate from the Jacobi-Eisfeld model and the alternative James-Birchall approach (Yeh-Schum lung geometry) are compared in Figure 2.28 for the three categories of room ventilation considered to represent houses. The range of doses shown in the figure corresponds with the variation of $f_p$ and AMD given in Figure 2.25, assuming that high values of $f_p$ are associated with small aerosols, and vice versa. This range includes extreme assumptions about mucus clearance rates and also retention characteristics (James-Birchall model).

Variations in the aerosol parameters, $f_p$ and AMD, for each ventilation category, give rise to a much larger dispersion in the dose conversion coefficient than differences between models or biological variables (c.f. Figure 2.27). Ventilation rate does not exclusively determine the regional lung dose for a given exposure, although there is a trend to lower dose conversion coefficients with low air change rates. This trend is not sufficient, however, to offset the higher exposures to potential $\alpha$-energy incurred as ventilation rate is reduced.

If an average breathing rate of 0.75 $m^3 h^{-1}$ is assumed for adults exposed indoors to $^{222}Rn$-daughers, the values of absorbed dose per unit of exposure to potential $\alpha$-energy (WLM and J h $m^{-3}$) given in Table 2.9 are derived as a function of the free atom fraction of potential $\alpha$-energy, $f_p$ and AMD of the attached aerosol.

It is seen from Table 2.9 that the differences between results of the two dosimetric models considered are much smaller than the variations in the conversion coefficients resulting from the physical variables; $f_p$ and AMD. The dose conversion coefficients derived from these results for the physical aerosol conditions assumed to typify different categories of room ventilation are given in Table 2.10. Results for both dosimetric models indicate that, per unit concentration of potential $\alpha$-energy in indoor air, the mean dose to the T-B region may be up to about a factor 2 higher in highly ventilated houses than in those with low air exchange rates.

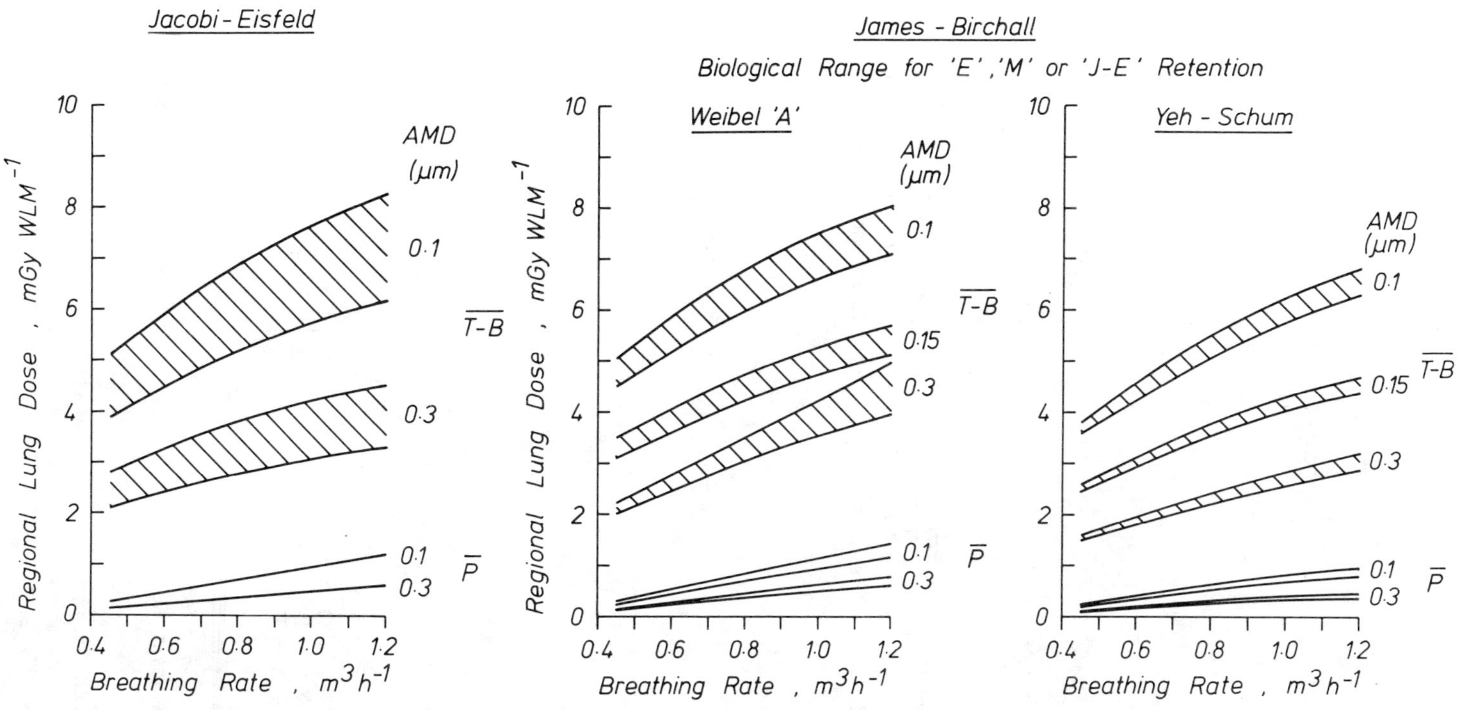


Figure 2.27 Regional lung dose ($\overline{T-B}$, $\overline{P}$) from 1 WLM exposure to $^{222}$Rn-daughters attached to aerosols as a function of breathing rate and AMD.

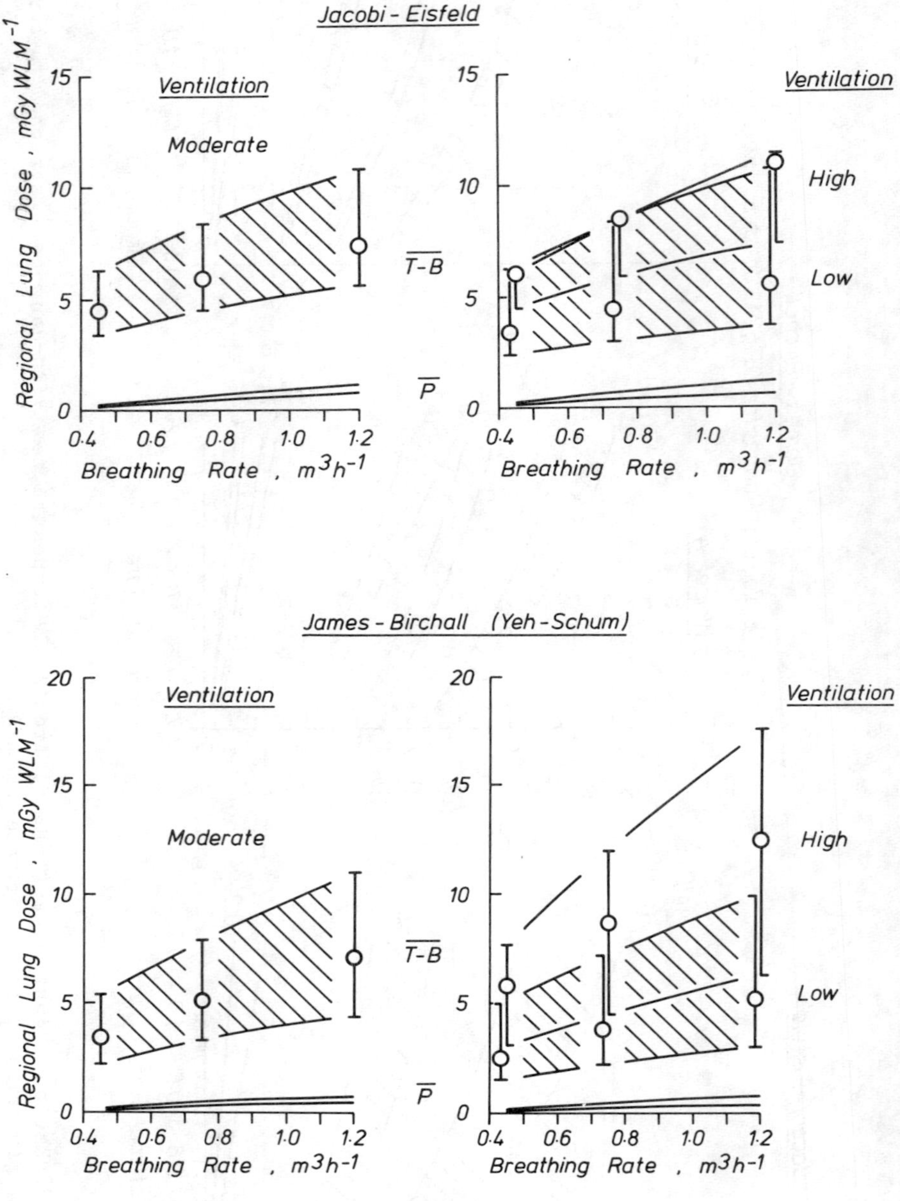


Figure 2.28 Variation of regional lung dose ($\overline{T-B}$, $\overline{P}$) with breathing rate and room ventilation.

Table 2.9

CONVERSION FUNCTIONS FOR EVALUATION OF THE REGIONAL LUNG DOSE, D AND EFFECTIVE DOSE EQUIVALENT, $H_E$ PER WLM EXPOSURE OF ADULT MEMBERS OF THE PUBLIC (NOSE BREATHING AT A MEAN RATE OF 0.75 $m^3$ $h^{-1}$) AND PER J h $m^{-3}$ EXPOSURE TO POTENTIAL $\alpha$-ENERGY(1,2,3)

| Conversion Function (1) | Dosimetric Model (2) | AMD = 0.1 μm | | AMD = 0.2 μm | |
|---|---|---|---|---|---|
| | | Lung Region | | Lung Region | |
| | | T-B | P | T-B | P |
| D(mGy) per WLM | J-E<br>J-B | $5.3+15f_p$<br>$5.0+62f_p$ | $1.3(1-f_p)$<br>$0.5(1-f_p)$ | $2.9+17fp$<br>$2.8+64f_p$ | $0.9\ (1-f_p)$<br>$0.3\ (1-f_p)$ |
| D(Gy) per J h $m^{-3}$ | J-E<br>J-B | $1.5+4f_p$<br>$1.4+18f_p$ | $0.4(1-f_p)$<br>$0.2(1-f_p)$ | $0.8+5f_p$<br>$0.8+18f_p$ | $0.3\ (1-f_p)$<br>$0.1\ (1-f_p)$ |
| $H_E$(mSv) per WLM | J-E<br>J-B | $7.9+15f_p$<br>$6.6+74f_p$ | | $4.5+19f_p$<br>$3.7+76f_p$ | |
| $H_E$ (Sv) per J h $m^{-3}$ | J-E<br>J-B | $2.3+5f_p$<br>$1.9+21f_p$ | | $1.3+6f_p$<br>$1.1+22f_p$ | |

(1) $f_p$ = unattached fraction of total potential $\alpha$-energy

(2) J-E = Dosimetric model of JACOBI-EISFELD with Weibel 'A' lung
J-B = Dosimetric model of JAMES-BIRCHALL with Yeh-Schum lung

(3) $H_E$ assumes quality factor, Q = 20 for $\alpha$-particles and weighting factor, $W_T$ = 0.06 for each regional lung dose

Table 2.10

DOSE CONVERSION COEFFICIENTS PER UNIT OF EXPOSURE TO POTENTIAL α-ENERGY $^{222}Rn$-DAUGHTERS (WLM) AT A MEAN ADULT BREATHING RATE OF 0.75 $m^3\ h^{-1}$ IN DOMESTIC ENVIRONMENTS AND PER $J\ h\ m^{-3}$ FOR DIFFERENT CATEGORIES OF ROOM VENTILATION(1,2)

| Ventilation Category | Model | D(mGy) per WLM T-B | D(mGy) per WLM P | D(Gy) per $J\ h\ m^{-3}$ T-B | D(Gy) per $J\ h\ m^{-3}$ P | $H_E$ (mSv) per WLM | $H_E$ (Sv) per $J\ h\ m^{-3}$ |
|---|---|---|---|---|---|---|---|
| Low | J-E | 3.2 | 0.8 | 0.9 | 0.2 | 4.9 | 1.4 |
| | J-B | 3.8 | 0.3 | 1.1 | 0.1 | 4.9 | 1.4 |
| Moderate | J-E | 4.5 | 1.1 | 1.3 | 0.3 | 6.6 | 1.9 |
| | J-B | 5.1 | 0.3 | 1.4 | 0.1 | 6.5 | 1.8 |
| High | J-E | 6.0 | 1.2 | 1.7 | 0.3 | 8.7 | 2.5 |
| | J-B | 8.7 | 0.4 | 2.5 | 0.1 | 10.9 | 3.1 |

(1) J-E = Dosimetric model of Jacobi-Eisfeld with Weibel 'A' lung
J-B = Dosimetric model of James-Birchall with Yeh-Schum lung

(2) The given values of dose per WLM exposure can be converted to annual dose per WL potential α-energy concentration, assuming an indoor occupancy factor of 0.8, by applying the coefficient:

$$1\ WL = 41\ WLM\ y^{-1}$$

### 2.5.4 Effective dose equivalent

Variation of the effective dose equivalent, $H_E$, per unit of exposure to potential α-energy with breathing rate and room ventilation is essentially the same as that of dose to the T-B region (Figure 2.28), since the bronchial dose is substantially higher than that to the pulmonary region. Average conversion coefficients between $H_E$ and exposure derived from the regional lung doses are given in Tables 2.9 and 2.10.

### 2.5.5 Reference dose conversion coefficients for adults exposed indoors

The ventilation and aerosol conditions in houses vary with time, both daily and seasonally. Variation of the radon daughter concentration with ventilation rate will be very much larger than, and in the opposite sense to, the small dependence of the dose conversion coefficients on aerosol conditions. Under these circumstances it is reasonable to apply constant dose conversion coefficients for the estimation of cumulative dose to individuals and collective dose to the whole population from inhaled radon daughters.

It is proposed that the average values of the dose conversion coefficients given in Table 2.10 for moderate-low ventilation rates will give the most appropriate assessment of lung doses to the general public, in view of the established trend in industrialised countries to reduce room ventilation in order to conserve energy. Reference dose conversion coefficients thus derived are given in Table 2.11.

Table 2.11

PROPOSED REFERENCE VALUES FOR ADULTS OF THE ABSORBED α-DOSE, D AND THE EFFECTIVE DOSE EQUIVALENT, $H_E$ PER UNIT OF

- potential α-energy concentration in air (in $Jm^{-3}$ or WL) per annum
- equilibrium equivalent $^{222}Rn$-concentration (in Bq $m^{-3}$) per annum
- intake of potential α-energy by inhalation (in Joules)
- exposure to potential α-energy (in J h $m^{-3}$ or WLM)

FOR EXPOSURE TO $^{222}Rn$-DAUGHTERS IN INDOOR ENVIRONMENTS

| Quantities and Units | | Regional Lung Dose | |
|---|---|---|---|
| | | T-B | P |
| Annual α-dose, D (Gy) | per $Jm^{-3}$ | 8000 | 1000 |
| | per WL | 0.16 | 0.02 |
| | per Bq ($^{222}Rn$) $m^{-3}$ | 45 x $10^{-6}$ | 5.6 x $10^{-6}$ |
| α-dose, D (Gy) | per J | 1.6 | 0.20 |
| | per J h $m^{-3}$ | 1.2 | 0.15 |
| | per WLM | 0.004 | 0.0005 |
| Annual effective dose equivalent, $H_E$ (Sv) | per J $m^{-3}$ | 11000 | |
| | per WL | 0.22 | |
| | per Bq ($^{222}Rn$) $m^{-3}$ | 60 x $10^{-6}$ | |
| Effective dose equivalent, $H_E$ (Sv) | per J | 2.0 | |
| | per J h $m^{-3}$ | 1.5 | |
| | per WLM | 0.0055 | |

(1) Calculated with $W_{T-B} = W_P = 0.06$ and $Q_\alpha = 20$

It must be noted that the reference values for the effective dose equivalent, $H_E$ of 2.0 Sv per Joule intake; 5.5 mSv per WLM exposure (1.5 Sv per J h $m^{-3}$) and the annual effective dose equivalents of 11000 Sv per J $m^{-3}$; 0.22 Sv per WL potential α-energy concentration in air, assume the same weighting factor of 0.06 for the sensitivity of bronchial and pulmonary tissue as that in miners. Any effect of exposure co-factors in mine air of enhancing the radiosensitivity of these tissues (section 4.6) may be less important in domestic environments. Hence the risk per unit of absorbed dose may well be lower in domestic environments than in mines, and the effective dose equivalents given in Table 2.11 could be regarded as conservative estimates for application to radiological protection.

### 2.5.6 Age dependence

#### Regional lung dose

The mean doses to the T-B region calculated as a function of age for 1 WLM exposure to $^{222}Rn$-daughters in moderately ventilated rooms are shown in Figure 2.29 for unattached atoms and Figure 2.30 for the aerosol fraction of potential $\alpha$-energy. Intake of potential $\alpha$-energy is assumed to be age dependent (section 2.2) and alternative assumptions are made about the depth distribution of basal cells: either that this is a function of age (f(a)), the more likely case, or that epithelial thickness is independent of age ($\neq$ f(a)).

Both dosimetric models considered give the same general age dependence of bronchial dose. The conversion coefficients calculated with the James-Birchall model for unattached atoms, however, are about an order of magnitude higher than those given by the Jacobi-Eisfeld calculation (Figure 2.29). This is due partly to the different biological retention assumed for free atoms in these two models, at all ages, but also to the high deposition in the trachea and main bronchi given by the Jacobi-Eisfeld model, especially in younger children (Figure 2.6). Dose to the trachea and main bronchi is excluded from the regional average.

There is a general tendency for bronchial dose from the unattached fraction of potential $\alpha$-energy to increase in younger children by about a factor of 1.5. This increase is due mainly to the assumed reduction in epithelial thickness with age, illustrated in Figure 2.29 by comparison with the lower doses given by assuming constant epithelial thickness.

Figure 2.30 shows that the age dependence of regional lung dose from the major fraction of the potential $\alpha$-energy attached to aerosol particles is also small, i.e. the regional dose is by about a factor of 1.5 higher in children than in adults. The two models give similar bronchial doses in this case.

Hofmann (1982) has calculated that dose to the bronchial region is about a factor of 2 to 3 higher in a 6 year old child than a 30 year old adult, depending on the assumptions of basal cell depths. This rather greater age dependence of dose than the above results arises principally from the different respiratory standards adopted by Hofmann and thus higher intake of potential $\alpha$-energy per unit exposure as a function of age.

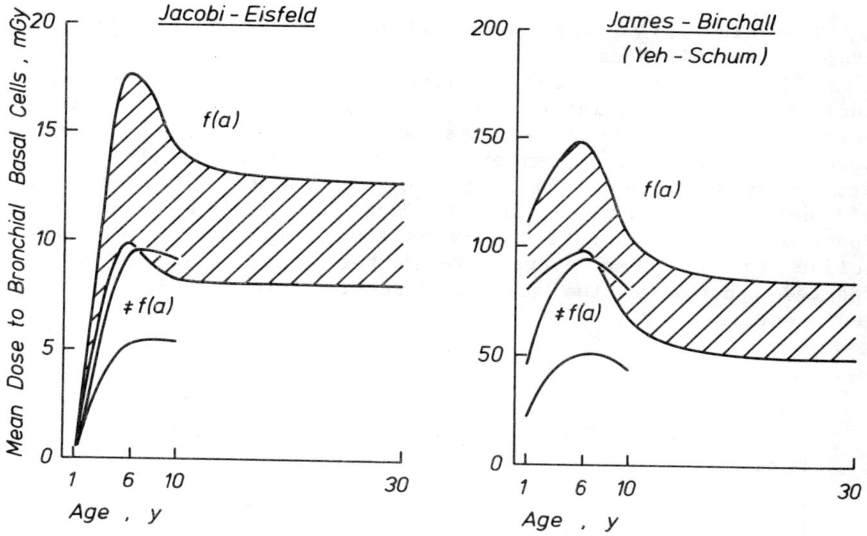


Figure 2.29 Age dependence of mean dose to bronchial basal cells for 1 WLM exposure to free $^{222}Rn$-daughter atoms ($f_p = 1$), assuming basal cell depth is a function of age (f(a)) or independent of age (≠ f(a)). Ranges correspond to biological variability incorporated in each model.

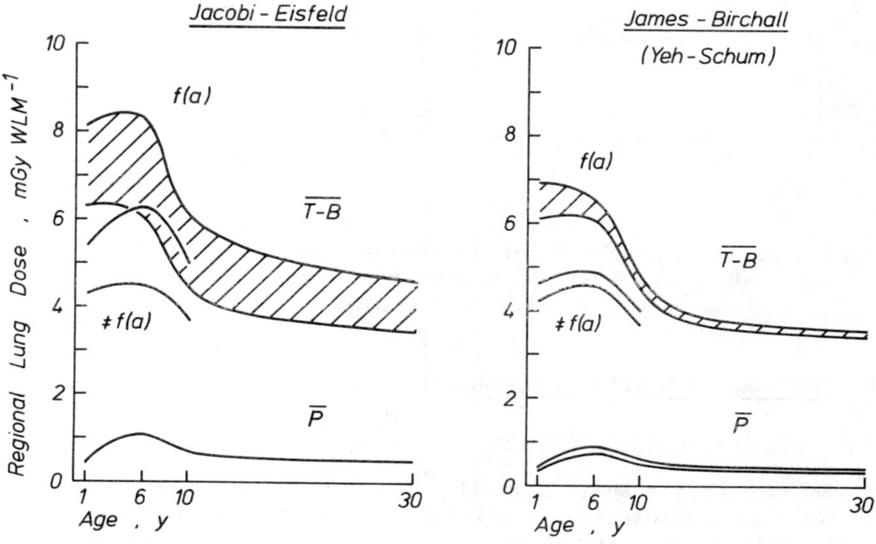


Figure 2.30 Age dependence of regional lung dose for 1 WLM exposure to 0.15 μm AMD aerosol ($f_p = 0$). Basal cell depths as in Figure 2.29.

Effective dose equivalent

The effective dose equivalent, $H_E$, as a function of age obtained by combining the contributions from the unattached fraction, $f_p$, and the attached aerosol characterising moderately ventilated rooms (Figure 2.25) is very similar for the two dosimetric models considered as shown in Figure 2.31. The ranges of conversion coefficient given in this figure include typical variations in $f_p$ and AMD about their mean values. Overall, the influence of age on the effective dose equivalent, $H_E$, per unit of exposure to $^{222}Rn$-daughters is less than a factor 2. Some further decline in the dose conversion factor after the age of 30 is expected because of the general reduction in breathing rates with age in adults.

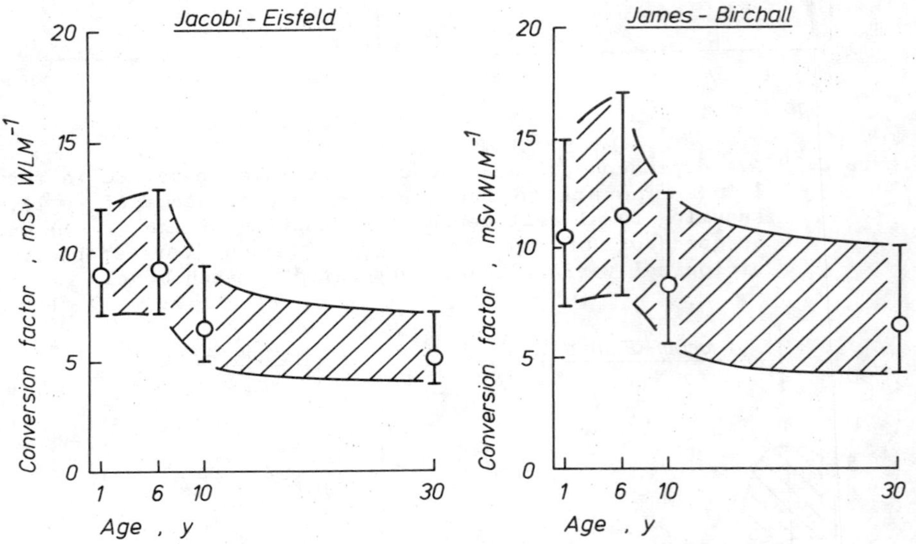


Figure 2.31 Age dependence of effective dose equivalent, $H_E$, for 1 WLM exposure to $^{222}Rn$-daughters in moderately ventilated rooms.

## 2.6 EXPOSURE TO RADON-220 DAUGHTERS

### 2.6.1 Aerosol Characteristics

Because of its short half-life (0.15 sec), airborne $^{216}Po$ (ThA) is close to radioactive equilibrium with thoron gas and exists almost entirely as unattached atoms or ions. The airborne concentration of the subsequent long-lived daughter $^{212}Pb$ (ThB, $T_{\frac{1}{2}}$ = 10.6 hr) is likely to be about 2 per cent of the equilibrium value. The fraction of $^{212}Pb$ remaining unattached to nuclei, $f_B$, is not well known, nor is the size distribution of the attached aerosol and its dependence on atmospheric conditions. Harley and Pasternack (1972a) and Hofmann et al (1979) have taken $f_B$ to be 0.02. Estimates of the AMD of the attached aerosol range from 0.08 μm in room air (Hofmann et al, 1979) to 0.2-0.3 μm in workplaces (Jacobi and Eisfeld, 1980).

It is reasonable, however, to expect the size distribution of the attached $^{220}$Rn-daughter aerosol to be larger than that of $^{222}$Rn-daughters in both workplaces and indoor air, since aerosols tend to increase in size with age. A likely mean value for the AMD of the $^{220}$Rn-daughter aerosol in indoor air may therefore be about 0.2 μm, as indicated by the experimental work of Martin and Jacobi (1972).

### 2.6.2 Regional lung dose

In the following sections the regional lung dose from exposure to potential α-energy of the $^{220}$Rn-daughters at a mean breathing rate of 1.2 $m^3$ $h^{-1}$, applicable to workplaces, is examined.

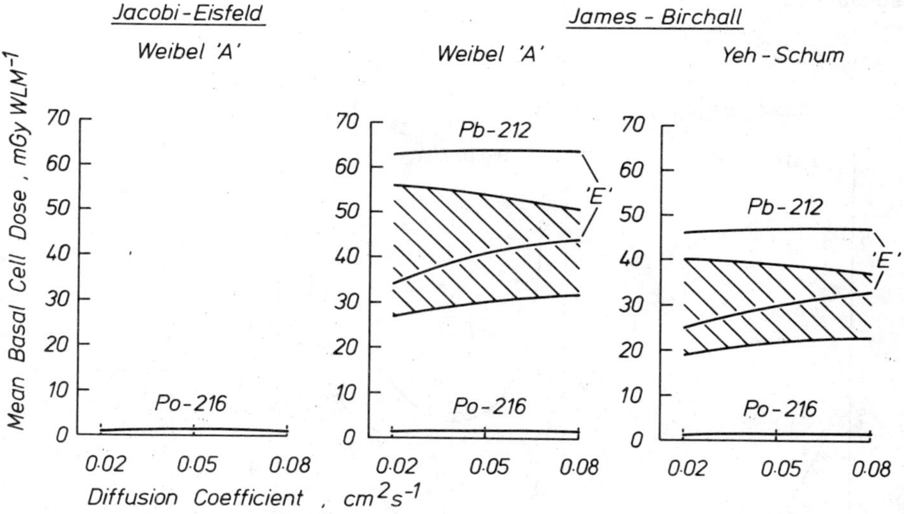


Figure 2.32 Mean dose to bronchial basal cells from 1 WLM exposure to free $^{220}$Rn-daughter atoms ($f_p$ = 1, 4.2 x $10^{-3}$ J intake) showing dependence on diffusion coefficient, ciliary transport rates and dosimetric model. Jacobi-Eisfeld model (rapid absorption) gives negligible dose for free $^{212}$Pb atoms. Exposure to $^{216}$Po assumed to be 0.02 WLM.

#### Unattached fraction

The calculated dose to the bronchial region depends strongly on the assumed biological retention of $^{212}$Pb atoms (Figure 2.32). With the Jacobi-Eisfeld model which assumed a biological half-time of 15 minutes, approximately 98% of the potential α-energy of unattached atoms associated with $^{212}Pb$ gives negligible bronchial dose. The dose from $^{216}Po$ (ThA) atoms is unaffected by clearance. For the James-Birchall model, bronchial dose varies in the range from about 20 to 60 mGy $WLM^{-1}$, depending on the assumed lung dimensions, site of retention at the bronchial surface and mucus transport-rates. The assumed value of the diffusion coefficient has a negligible effect on bronchial dose from $^{220}Rn$-daughter atoms.

Attached Aerosol

The assumed characteristics of biological retention have a larger influence on the mean dose to the bronchial region from $^{220}Rn$-daughters than aerosol size (Figure 2.33). According to the J-E model, when retention is determined by the rate of desorption from carrier particles with a half-time of about 8 hours followed by rapid absorption into the blood, bronchial dose is reduced by about a factor 2. Gross redistribution of deposited $^{212}Pb$ by ciliary clearance leads to a substantial dependence of bronchial dose on the assumed transport rates.

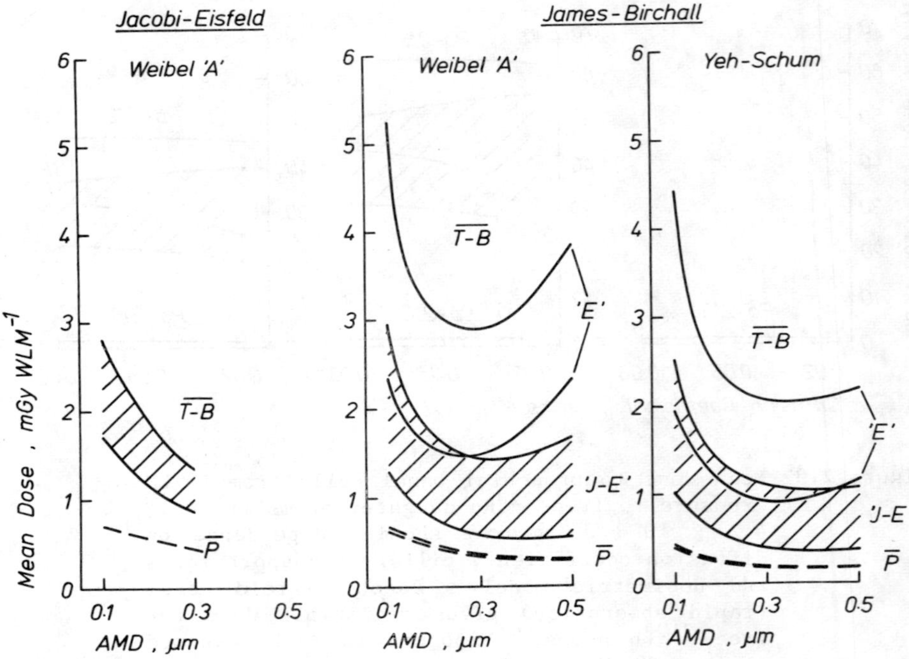


Figure 2.33 Regional lung dose ($\overline{T\text{-}B}$, $\overline{P}$) as a function of AMD of attached $^{220}Rn$-daughter aerosol and dosimetric model. Bronchial doses for retention in mucus (James-Birchall model) are 5-10% lower than values for retention in epithelium 'E'.

Irrespective of the deposition and retention model used, dose to the bronchial region is reduced by about a factor 2 over the aerosol size range 0.1 to 0.3 μm AMD. The additional contribution to bronchial dose from approximately 2% of the potential α-energy associated with unattached atoms amounts to about 20% assuming long-term retention of $^{212}Pb$, or zero assuming the desorption model of Jacobi-Eisfeld.

Summary of dose conversion coefficients

Average values of the conversion coefficients between exposure to potential α-energy and dose estimated by applying the various deposition and clearance models are given in Table 2.12 in terms of the variable physical characteristics of the inhaled $^{220}Rn$-daughter aerosol. Results for the J-B model vary by about ± 50% depending on the assumptions of biological clearance.

Taking into account the uncertainties in the size distribution of the attached aerosol and biological retention, the mean bronchial dose from $^{220}Rn$-daughters ($^{212}Pb$ and $^{212}Bi$) probably lies in the range 1-2 mGy $WLM^{-1}$ (0.3-0.6 Gy per J h $m^{-3}$) and pulmonary dose 0.2-0.5 mGy $WLM^{-1}$ (0.06-0.15 per J h $m^{-3}$).

Table 2.12

CONVERSION FUNCTIONS FOR EVALUATION OF THE REGIONAL LUNG DOSE, D PER WLM EXPOSURE TO $^{220}Rn$-DAUGHTER POTENTIAL α-ENERGY (AT A MEAN BREATHING RATE OF 1.2 $m^3$ $h^{-1}$) AND PER J h $m^{-3}$ (1,2,3)

| Conversion Function (1) | Dosimetric Model (2) | AMD = 0.2 μm | | AMD = 0.3 μm | |
|---|---|---|---|---|---|
| | | Lung Region | | Lung Region | |
| | | T-B | P | T-B | P |
| D(mGy) per WLM | J-E | 1.7 | $0.6\ (1-f_p)$ | 1.3 | $0.4\ (1-f_p)$ |
| | J-B | $1.5+29f_p$ [2.1] | $0.3\ (1-f_p)$ | $1.3+29f_p$ [1.9] | $0.2\ (1-f_p)$ |
| D(Gy) per J h $m^{-3}$ | J-E | 0.5 | $0.17(1-f_p)$ | 0.36 | $0.12(1-f_p)$ |
| | J-B | $0.43+8f_p$ [0.6] | $0.08(1-f_p)$ | $0.36+8f_p$ [0.5] | $0.06(1-f_p)$ |

(1) $f_p$ = unattached fraction of potential α-energy ($f_p = f_B \approx 0.02$)

(2) J-E = Dosimetric model of JACOBI-EISFELD with Weibel 'A' lung
J-B = Dosimetric model of JAMES-BIRCHALL with Yeh-Schum lung
[ ] = Bronchial dose given by J-B model assuming $f_p$ = 0.02

(3) Activity ratio $^{212}Bi$: $^{212}Pb$ assumed to be 0.25: 1. Intake of $^{212}Pb$ contributes approx. 98% of the total potential α-energy

### 2.6.3 Effective Dose Equivalent

A significant fraction of deposited $^{212}Pb$ (ThB) is transferred from the pulmonary region to other organs of the body via the blood (Jacobi and Eisfeld, 1980; 1981). The tissue dose equivalents that result for aerosols in the size-range 0.2 - 0.3 μm AMD, taking into account uncertainty in biological retention, are given in Table 2.13.

Table 2.13

TISSUE DOSE EQUIVALENTS AND EFFECTIVE DOSE EQUIVALENTS PER UNIT OF EXPOSURE TO POTENTIAL α-ENERGY $^{220}Rn$-DAUGHTERS AT A MEAN BREATHING RATE OF 1.2 $m^3\ h^{-1}$, FOR AEROSOL AMD IN THE RANGE 0.3 - 0.2 μm(1)

| Target Tissue | Tissue Dose Equivalent per unit exposure (Sv per J h $m^{-3}$) | Effective Dose Equivalent per unit exposure (Sv per J h $m^{-3}$) |
|---|---|---|
| T-B region | 6 - 11 | 0.4 - 0.7 |
| P region | 1.2 - 2.4 | 0.07 - 0.14 |
| Bone Surfaces | 2.4 - 5 | 0.07 - 0.13 |
| Red Bone Marrow | 0.18 - 0.4 | 0.02 - 0.04 |
| Kidneys | 1.4 - 3 | 0.08 - 0.18 |
| Liver | 0.3 - 0.6 | 0.02 - 0.04 |
| Spleen | 0.06 - 0.12 | - |
| Other Tissues | 0.02 - 0.05 | - |
| Total | - | 0.6 - 1.2 |

1) Dose per unit exposure increases as aerosol size decreases

The effective dose equivalent for inhalation of $^{220}Rn$ and its daughters, given by summing the contributions from each irradiated tissue weighted according to radiosensitivity (ICRP, 1977) is estimated to be between 0.6 and 1.2 per J h $m^{-3}$ (approximately 2-4 mSv $WLM^{-1}$ exposure at a breathing rate of 1.2 $m^3\ h^{-1}$) for aerosol AMD in the range 0.3-0.2 μm. Approximately half of this effective dose equivalent is contributed by irradiation of bronchial tissue.

### 2.6.4 Influence of breathing rate

It is reasonable to consider the tissue dose equivalents and effective dose equivalents per unit intake of $^{220}Rn$-daughter potential α-energy to be independent of breathing rate over the range appropriate to occupational and domestic exposure of adults (1.2 $m^3\ h^{-1}$ to 0.45 $m^3\ h^{-1}$). Hence for adults, dose conversion coefficients per unit exposure to potential α-energy (Table 2.13) can be taken as proportional to breathing rate.

## 2.7 CONCLUSIONS

Bronchial stem cells are considered to be the cells at risk for induction of bronchogenic lung cancer by inhalation of radon daughters. On the basis of this assumption the bronchial basal cell layer is the relevant target tissue for lung dosimetry of inhaled short-lived radon daughters. It follows from dosimetric analysis that the mean dose to the basal cell layer of the ciliated bronchial epithelium is a factor 3-10 higher than that to the alveolar epithelium and the terminal non-ciliated bronchioles.

Sensitivity analysis indicates that the conversion factor between intake of potential α-energy and absorbed dose depends on the unattached fraction, $f_p$, of the total potential α-energy of the radon daughter mixture and on the activity median diameter (AMD) of the carrier aerosol for the attached daughter atoms. This conversion factor is independent of the state of radioactive equilibrium between the individual radon daughters. For relevant values of the aerosol parameters the results of the two dosimetric studies are in good agreement, although different models of lung geometry, aerosol deposition, lung retention and the depth distribution of bronchial stem cells have been considered.

On the basis of these findings reference values for the conversion factors between exposure to potential α-energy and the α-dose to target tissues in the lung from inhaled radon daughters can be derived.

In the following these dose factors for occupational exposure in mines and for population exposure in houses are summarized.

The results are not only of importance with respect to lung dosimetry; they also indicate the quantities which should be measured in air monitoring programmes.

In addition the derived dose factors may be useful for analysis of the lung cancer risk associated with inhalation of radon daughters. It should be emphasized, however, that dosimetric models cannot describe the possible influence of co-carcinogenic or synergistic factors on the induction and promotion of lung cancer. This is particularly valid for the conditions in mines. It is likely that such factors have contributed to the excess lung cancer risk observed among miners exposed to $^{222}$radon daughters (ICRP, 1981).

### 2.7.1 Radon-222 Daughters in mines

The conversion factor between exposure to potential α-energy and the mean dose to bronchial stem cells is relatively insensitive to biological uncertainties in modelling. Some variation of this factor is expected with the physical characteristics of $^{222}$Rn-daughter aerosols under different atmospheric conditions in mines.

In underground mines aerosol characteristics are expected to vary over a range of conditions from poor ventilation with high dust loading, large aerosol particles and low values of $f_p$; giving a mean bronchial dose of about 3 mGy $WLM^{-1}$ (0.12 Gy per J h $m^{-3}$), to high ventilation, clean air, small aerosol particles and relatively high $f_p$; giving rise to conversion factors in the range 7-10 mGy $WLM^{-1}$(2.2-3 Gy per J h $m^{-3}$).

Lifetime exposure in underground mines may be typified by moderate conditions of ventilation and dust generation, with $f_p$ about 0.02. Under these conditions dose conversion coefficients approximately 6 mGy $WLM^{-1}$ (1.8 Gy per J h $m^{-3}$) are derived, depending little on the dosimetric model adopted for the calculation. The corresponding range of doses to bronchial stem cells calculated for long-term exposure in open pit mines is about 5-8 mGy $WLM^{-1}$ (1.4-2.4 Gy per J h $m^{-3}$). A range of dose conversion factors about twice these values is calculated for short-term exposure under more extreme environmental conditions in open-pit mines ($f_p \sim 0.1$ and AMD of attached aerosol ~ 0.1 μm).

Dose to pulmonary tissue is calculated to be a factor 5-10 less than the mean dose to bronchial stem cells, depending on the dosimetric model adopted. A range of about 0.5-1.2 mGy $WLM^{-1}$ (0.12-0.3 Gy per J h $m^{-3}$) is derived.

In order to limit dose from occupational exposure to $^{222}Rn$ daughters it is necessary to consider a conversion factor between exposure to potential α-energy and effective dose equivalent, $H_E$, which is determined predominantly by the mean dose to bronchial stem cells. It is calculated that $H_E$ varies with aerosol characteristics in underground mines in the range of about 6-10 mSv $WLM^{-1}$ (1.8-3 Sv J h $m^{-3}$). A reference value of 8.5 mSv $WLM^{-1}$ (2.4 Sv per J h $m^{-3}$) is proposed to characterise lifetime exposure to $^{222}Rn$-daughters in all mines.

Radiobiological interpretation of the observed incidence of cancers in the upper part of the bronchial tree and the influence of chemical exposure factors in mine air requires additional knowledge of the distribution of dose in the upper airways. Estimates of dose to basal cells in segmental bronchi are strongly dependent on physical aerosol parameters, notably $f_p$, but they are also influenced by biological uncertainty in modelling to the extent of about a factor 3.

A major source of uncertainty in the assessment of dose to the upper bronchial tree is the limited available information on the depths of sensitive basal cells below the airway surface. It is considered that basal cells in segmental bronchi lie deeper in the epithelium than assumed in earlier calculations. Hence these cells are now considered to receive about the same or, depending on the model adopted, rather less dose than the average for the bronchial region as a whole.

The homogeneity in calculated dose to bronchial stem cells, when compared with the reported preponderance of cancers in the segmental bronchi, indicates that bronchial tissue in miners may not be uniformly sensitive to α-irradiation. Any modifying effect of exposure to chemical co-factors in mine air on the radiosensitivity of bronchial tissue is expected to be expressed predominantly in the upper airways, particularly the segmental bronchi. Hence exposure co-factors may be important in determining the risk of bronchial irradiation by $^{222}$Rn-daughters.

2.7.2 Radon-222 daughters in houses

The rate of exchange between indoor and outdoor air is expected to influence the free atom fraction of potential α-energy, $f_p$ and the activity median diameter, AMD of the attached aerosol. Ventilation, however, has a much greater effect in reducing the concentration of $^{222}$Rn-daughters in indoor atmospheres.

The range of $f_p$ typical of domestic atmospheres is higher than that in underground mines and the AMD smaller. Both of these factors increase the mean dose to bronchial basal cells for per unit exposure to potential α-energy.

Intake of potential α-energy for a given exposure, however, is lower in domestic environments than in mines; because of the lower mean breathing rate applicable during the time spent in the home.

According to models based on the regional lung dose concept, breathing rate has a greater overall effect on dose than differences in physical aerosol characteristics between mines and homes; leading to lower average lung doses in homes per unit exposure to potential α-energy.

The mean dose to bronchial basal cells of adults exposed indoors is calculated to lie in the range 3-9 mGy per WLM exposure (0.9 - 2.5 Gy per J h $m^{-3}$), increasing with the assumed ventilation rate, but essentially independently of the choice of dosimetric model. Pulmonary doses in the range 0.3 - 1.2 mGy $WLM^{-1}$ (0.08 - 0.3 Gy per J h $m^{-3}$) are calculated, depending both on dosimetric model and aerosol characteristics.

It is proposed for the purpose of radiological protection that aerosol characteristics typical of rooms ventilated at a rate substantially less than 1 air change per hour are assumed. Reference conversion coefficients of 4 mGy per WLM exposure (1.2 Gy per J h $m^{-3}$) for the mean bronchial dose and 0.5 mGy $WLM^{-1}$ (0.15 Gy per J h $m^{-3}$) for pulmonary dose are thus derived.

These regional lung doses give an effective dose equivalent of 5.5 mSv per WLM exposure (1.5 Sv per J h $m^{-3}$) if the radiosensitivity of lung tissue in adult members of the public is assumed to be the same as that in miners exposed to more toxic atmospheres. This procedure may conservatively over-estimate the risk of exposure to $^{222}$Rn-daughters in homes.

Averaged over the whole age period from birth to 10 years it is reasonable to consider the effective dose equivalent per unit exposure to be about a factor 1.5 higher than the reference adult value. On the other hand, for old people a correction factor somewhat less than 1 should be expected, taking into account the decrease in breathing rate with age. For evaluation of the total lifetime dose, therefore, the reference conversion factor derived for the 30 year old adult can be applied without any age correction.

### 2.7.3 Radon-220 Daughters

The physical characteristics of $^{220}$Rn-daughter aerosols are less well known than those of $^{222}$Rn-daughters. It is likely, however, that aerosol size is generally larger but less variable than that of $^{222}$Rn-daughters and the unattached fraction of potential $\alpha$-energy lower.

Uncertainty in the biological parameters required for dosimetric modelling gives rise to a range of about a factor 2 in estimates of mean bronchial dose. The results of dosimetric modelling are less sensitive to variations in aerosol characteristics expected between different environments, whether occupational or domestic.

Calculated Mean doses to bronchial stem cells lie in the range 1 -2 mGy per WLM exposure at a mean occupational breathing rate of 1.2 $m^3$ $h^{-1}$ (0.3 - 0.6 Gy per J h $m^{-3}$) for $^{220}$Rn-daughter aerosols in the probable size range 0.3 - 0.2 µm AMD. Essentially all of this dose arises from intake of $^{212}$Pb (thorium-B). The corresponding mean pulmonary doses are 0.2 - 0.6 mGy per WLM exposure (0.06 - 0.17 Gy per J h $m^{-3}$).

Absorption of $^{212}$Pb through the alveolar epithelium into the blood and subsequent uptake in other tissues, principally on bone surfaces and in the kidneys, contributes approximately equally with lung irradiation to the effective dose equivalent.

A range of effective dose equivalents from 2-4 mSv per WLM exposure (0.6 -1.2 Sv per J h $m^{-3}$) is derived for occupational exposure, the reference value depending principally on the assumed dosimetric model. The effective dose equivalent per unit exposure to $^{220}$Rn-daughter potential $\alpha$-energy is thus approximately 1/3 that for the $^{222}$Rn-daughters.

In the case of domestic exposure of adults to $^{220}$Rn-daughters it is reasonable to consider the effective dose equivalent per unit intake of potential $\alpha$-energy to be the same as that under occupational conditions. Hence the dose per unit of exposure can be taken as proportional to the mean breathing rate.

## Chapter 3

## ADEQUACY OF EXPOSURE MEASUREMENT AS AN INDEX OF DOSE

### 3.1 INTRODUCTION

It is common practice to estimate doses following intakes of radionuclides from measurements of their uptake and retention in the various organs and tissues (ICRP, 1969) of the exposed individual. In some situations, either because levels of radioactivity are too low to measure in vivo, radioactive half-lives too short, or when large numbers of people are involved, doses are estimated from measured concentrations in air, water, food etc., estimated breathing and/or ingestion rates, and estimated average uptakes and retentions based on some "reference" values (cf ICRP, 1975; ICRP, 1979). The estimation of doses from the inhalation of radon and/or thoron daughters is a special case of the latter. Because of their short half-lives and highly non-uniform distribution in the lung, it is essentially impossible to measure distribution and retention of the deposited daughters. Hence estimates of doses from exposures to radon and thoron daughters are based on measured air concentrations and relationships such as those derived in Chapter 2 between exposure to radon and thoron daughters and the average dose to the stem cells of the bronchial epithelium and to other tissues for various aerosol conditions, ages, and breathing rates.

There is considerable uncertainty in the relationships between exposure (the Working Level or potential alpha energy concentration of radon/thoron daughters in air) and these doses, as is discussed in Chapter 2. The causes of this uncertainty can be divided into two groups; firstly imprecision of our knowledge of such factors as deposition patterns, clearance rates and source-target geometry, and secondly such factors as aerosol particle size, the fraction of daughter atoms not attached to particles, breathing rates and age. The first group of factors are those that cannot be measured directly as part of a monitoring program, while the second group, at least in theory, can be.

This Chapter asks the question "Is it sufficient to measure the exposure to radon/thoron daughters or should other factors also be measured to insure that the dose can be estimated with sufficient accuracy?" In order to answer this question, criteria on what is sufficient accuracy (or criteria for adequacy of the WL or potential alpha energy concentration) are developed. The results of the model calculations of Chapter 2 are summarized and the variations of these results with the measurable factors are compared to the criteria for adequacy to give limits on these factors within which only the exposure to radon or thoron daughters need be measured.

The relationships given in Chapter 2 use the units WL (working level) and J $m^{-3}$ for the concentration of potential alpha energy in air, WLM (working level month), J h $m^{-3}$ for exposure, and J for inhaled potential alpha energy. These units and the relationship between them are summarized below for convenience (see also Appendix A).

The WL, while historically related to the equilibrium concentration (Holaday, 1969) of the short-lived daughters of $^{222}Rn$ (Radon) in one litre of air, is now more often related only to the potential alpha energy from decay of the short-lived daughters to $^{210}Pb$. The WL for Radon is thus defined as any activity of $^{218}Po$ (RaA), $^{214}Pb$ (RaB), $^{214}Bi$ (RaC) and $^{214}Po$ (RaC') in one litre of air that will result in $1.3 \times 10^5$ MeV of alpha energy in their decay to $^{210}Pb$. Similarly, the WL for $^{220}Rn$ (Thoron) is defined as the activity of $^{216}Po$ (ThA), $^{212}Pb$ (ThB), $^{212}Bi$ (ThC), $^{212}Po$ (ThC') or $^{208}Tl$ (ThC") in one litre of air that will result in $1.3 \times 10^5$ MeV of alpha energy in their decay to $^{208}Pb$.

The working level month (WLM) is defined as $170 \times 1.3 \times 10^5$ MeV h $l^{-1}$. That is, the exposure at an exposure rate of one WL ($1.3 \times 10^5$ MeV $l^{-1}$) for 170 h (one working month) would be one WLM. The SI equivalents of these units are: J $m^{-3} \cong 4.80 \times 10^4$ WL and J h $m^{-3} \cong$ 285 WLM.

The rate of intake of potential alpha energy into the lung when breathing air containing a radon daughter concentration expressed in WL units at a rate of B $m^{-3}$ $h^{-1}$ is

$$\dot{I}_p = 2.08 \times 10^{-5} \text{ B WL} \quad (\text{J h}^{-1}) \quad ....... \quad (3.1)$$

and the total intake of potential alpha energy for an average breathing rate $\bar{B}$ $m^3$ $h^{-1}$ is

$$I_p = 3.5 \times 10^{-3} \ \bar{B} \text{ WLM} \quad (\text{J}) \quad ....... \quad (3.2)$$

In the rest of this Chapter, the unit used for the exposure ($E_p$) is J h $m^{-3}$. Criteria are developed for the accuracy requirement in the relationship between this quantity and the dose. Limits on various parameters that affect this relationship are given between which the exposure is adequately correlated with dose.

## 3.2 CRITERIA FOR ADEQUACY

Criteria developed for the required accuracy and precision of a particular measurement or calculation must consider a minimum of three points:

i) The end use of the results

Two such end uses have been identified. The first is the limitation of dose associated with exposures to radon and/or thoron daughters for the purpose of radiation protection. The second is the unbiased estimate of the actual dose associated with exposure to radon and/or thoron daughters for the purposes of radiobiology and epidemiology. Because of the different nature of these two end uses, separate criteria should be developed for them. The limitation of dose for radiation protection is relatively straight forward, and criteria for this purpose, based on the effective dose equivalent, are developed below. Criteria for the purpose of epidemiological and radiological studies are not so straight forward, as the dose distribution in the bronchial tree may be important and not just the mean regional dose. Criteria for the accuracy of the relationship between exposure and the mean dose to bronchial stem cells can be developed in the same way as those for radiation protection.

ii) The precision and accuracy of the other measurements and/or calculations that go into obtaining the final result

It is well known (Johnson, 1978; McGregor, et al, 1980) that the calculation of exposures from the commonly used "grab sample" procedure has large uncertainties, primarily from the fluctuation in concentrations. Personnel exposure monitors and/or integrated area monitors have the potential of considerably reducing these uncertainties. However, the use of these monitors is not wide spread. In addition, bias in the estimated exposure may result from the sampled air not being representative of the breathing air, or, particularly in the case of personnel exposure monitors, from malfunctions or loss of calibration of the monitor or read-out device.

Another area of large uncertainty is caused by individual variability. It is possible to calculate accurately the dose to the various generations of the bronchial tree of a model lung provided the required aerosol measurements are made. However, individual variability in breathing patterns and clearances will still result in considerable uncertainty in doses to individuals. The magnitude of the uncertainties resulting from some of these variables is described in Chapter 2.

A final source of large uncertainty, relevant to radiobiology and epidemiology is the uncertainty in the number of observed lung cancers (or other effects) attributable to the radon or thoron daughters, due to their small numbers, the selection of an appropriate control group, the possibility of unknown synergistic or cofactors, and the possible variation in sensitivity between various groups which are otherwise identical.

It can be seen then that the lack of precision in the relationship between exposure and dose is only one of several uncertainties, and it would not be productive to demand too high a degree of accuracy in this relationship.

iii) The cost of improving the measurement and/or calculation

Continuous monitoring of aerosol parameters (mainly unattached fractions, and particle size) in breathing air would greatly improve our accuracy in calculating the doses to model lungs from radon and/or thoron exposures. However, these measurements would be costly, and particularly for radiation protection may not be justified at the low average exposure rates currently found in uranium and other mines. In this context, it might be useful to try to apply the ALARA (as low as reasonably achievable) principle. If it is argued that investigation or monitoring of aerosol parameters might require a further reduction of exposure in pursuit of optimisation, the cost of making these additional measurements should be compared with the benefit of dose reduction as part of the optimisation process.

### 3.2.1 Radiation protection criteria

There is no internationally accepted criterion for the accuracy required for internal dose calculations (which include the accuracy in dosimetric and metabolic models and in monitoring results), of which radon and thoron daughter dosimetry is a special category. The ICRP (ICRP, 1968) gives criteria for external exposure, and these have been applied previously to internal dosimetry (Johnson, 1976; HWC, 1980). The IAEA (IAEA, 1980) states for external dose assessment, that "The accuracy of the dose equivalent assessment should be appropriate to the level of exposure", and for internal dose states that "the required accuracy" may have to be taken into account when assessing dose equivalent, but does not state what that requirement is, or who will require it.

The ICRP's recommendation (ICRP, 1968) on the accuracy of external dose assessment for occupational exposure will be used here as guidance. It states (paragraph 101)

> The accuracy required in routine monitoring
>
> (101) The uncertainties acceptable in routine individual monitoring should be somewhat less than the investigation level and can best be expressed in relation to the annual dose. The uncertainty in assessing the upper limits to the annual dose equivalent to the whole body or to the organs of the body (see paragraph 92) should not exceed 50%. Where these doses are less than 2 rems an uncertainty of 1 rem is acceptable. This uncertainty includes errors due to variations in the dosemeter sensitivity with incident energy and direction of incidence, as well as intrinsic errors in the dosemeter and its calibration.

Three points on these recommendations need to be made:

i) The criterion is for the assessment of the upper limit on dose, and not the best estimate of dose. That is, ICRP recommends that the estimate of the dose be not less than 50% of the upper confidence limit on dose. This restriction could lead to a biased estimate of the dose. The upper limit has been taken to be the 95% confidence limit (Johnson, 1976).

ii) Since we are concerned here with the relationship between exposure and dose, and not absolute dose, the problem of accuracy at low doses, or exposure, does not concern us. This of course, would be of concern in instrument design and operation. We should not, however, be too concerned about the accuracy of the dose from low thoron exposures if radon exposures are high, and vice versa.

iii) The accuracy requirement includes the accuracy of the overall dose assessment, which in the case of radon and/or thoron daughters would include the uncertainty in the estimation of exposure and the uncertainty resulting from individual variability, as well as that resulting from the conversion of exposure to dose in model lungs, for which we are attempting to develop criteria.

Based on the above considerations, we shall label the exposure measurement adequate (that is, no other measurements are required for radiation protection purposes) for a given set of conditions if the factor converting exposure to effective dose equivalent, as calculated for the J-E and J-B models in Chapter 2 is not greater than 1.5 times the conversion factor implied in ICRP Publication 32 (ICRP, 1981). We shall also examine the adequacy of exposure as an index of dose should a higher annual limit of exposure be adopted on the basis of the conversion factor recommended in Chapter 2 to represent 'average' conditions in mines.

Occupational exposure

The ICRP has recommended, (ICRP, 1981) based on a review of the available data from epidemiological and dosimetric studies, that the annual limit on occupational exposure be 0.017 J h $m^{-3}$ for radon daughters and 0.050 J h $m^{-3}$ for thoron daughters. If these exposures are assumed to be equal to the ICRP recommended annual limit on effective dose equivalent of 0.05 Sv, the implied conversion factor is approximately 3.0 Sv per J h $m^{-3}$ and 1.0 Sv per J h $m^{-3}$ for radon and thoron daughters, respectively.

The dosimetry review given in Chapter 2 resulted in a proposed reference value for radon exposure to miners (see Table 2.8), based on average conditions that are assumed to represent long-term exposure in mines, of 8.5 mSv per WLM, or 2.4 Sv per J $m^{-3}$ h. It was also found that for a fixed breathing rate of 1.2 $m^3$ $h^{-1}$, the conversion factor only depended on particle size and unattached fraction to any significant degree. On the basis of this dosimetric approach, without also considering epidemiological data, an annual limit on occupational exposure of 0.021 J h $m^{-3}$ could be justified.

For occupational exposure to thoron a range in the value of effective dose equivalent per unit exposure of 0.6 to 1.2 Sv per J h $m^{-3}$ is derived in Section 6 of Chapter 2. However, no reference values are given because of the lack of adequate particle size and unattached fraction measurements upon which they can be based.

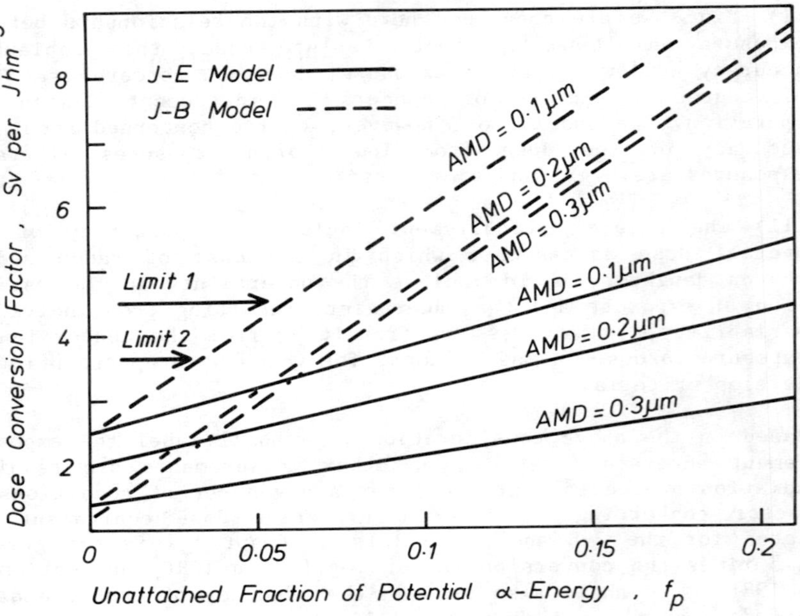


Figure 3.1 Factors for converting $^{222}$Rn-daughter exposure to dose as a function of unattached fraction for several aerosol sizes using the J-E and J-B models. Limit 1 for the conversion factor is 1.5 times the value recommended in ICRP Publ. 32 and relates to an annual exposure limit of 0.017 J h $m^{-3}$. Limit 2 is 1.5 times the reference value given in Chapter 2 for typical occupational exposure and relates to a higher annual limit on exposure thus derived of 0.021 J h $m^{-3}$.

Following the criteria developed above, the measurement of radon daughter exposure will be an adequate method of limiting dose in occupational exposure if the conversion factor does not exceed 4.5 Sv per J h $m^{-3}$ in relation to the ICRP exposure limit of 0.017 J h $m^{-3}$, and 3.6 Sv per J h $m^{-3}$ in relation to an exposure limit of 0.021 J h $m^{-3}$ based on the reference value from Chapter 2. These "limiting conversion factors" are shown on Figure 3.1, which summarizes the results of Chapter 2 for adults breathing at 1.2 $m^3$ $h^{-1}$ (Table 2.6). The measurement of thoron daughter exposure will be an adequate method of limiting doses provided the conversion factor does not exceed 1.5 Sv per J h $m^{-3}$ in relation to the ICRP exposure limit of 0.05 J h $m^{-3}$.

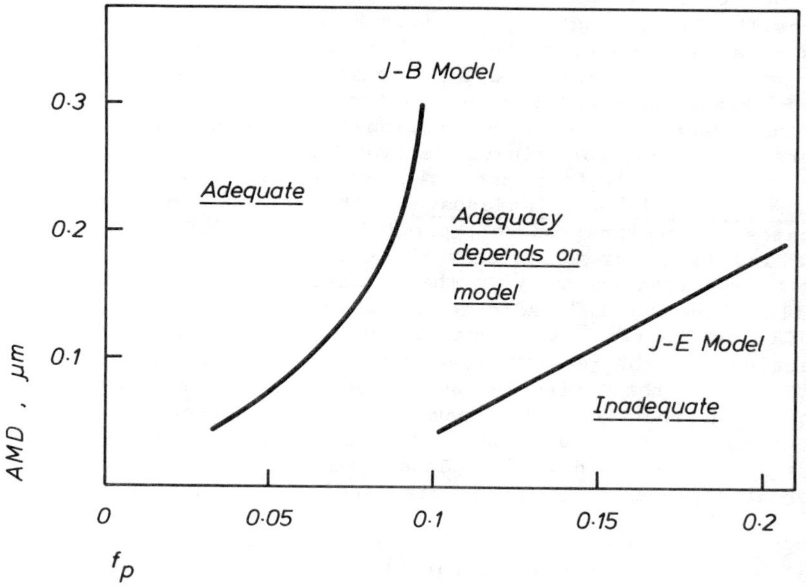


Figure 3.2 Estimated values of aerosol size and unattached fraction at which the conversion factor from Figure 3.1 exceeds Limit 1 based on value recommended in ICRP Publ. 32.

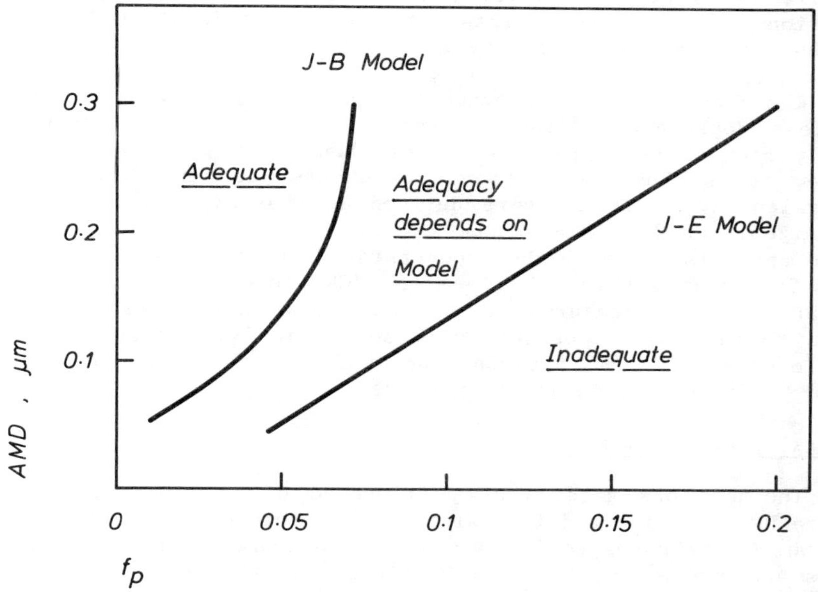


Figure 3.3 Estimated values of aerosol size and unattached fraction at which the conversion factor from Figure 3.1 exceeds Limit 2 based on reference value for occupational exposure given in Chapter 2.

The points at which the conversion factor for radon daughters exceeds the limits can be taken from Figure 3.1 for various particle sizes as a function of the unattached fraction. These points can be drawn on a plot of particle size against unattached fraction as is done in Figure 3.2 and Figure 3.3 for exposure limits based on the ICRP reference value for occupational exposure and the reference conversion factor for miners derived in Chapter 2, respectively. These plots are divided into regions labelled Adequate, Adequacy depends on Model, and Inadequate. The region labelled "adequate" indicates a combination of particle size (AMD) and unattached fractions ($f_p$) for which only the exposure needs to be measured to meet the required accuracy in the estimation of dose. Those regions labelled "inadequate" indicate a combination of AMD and $f_p$ for which the ICRP or Chapter 2 reference value for Figure 3.2 and Figure 3.3 respectively would result in an estimate of dose which was too low according to the criteria developed above, and those labelled "adequacy depends on model" would meet the criteria if conversion factors based on the J-E model were used, but not if the J-B model is used. This model dependency results primarily from their different dependence on $f_p$, as discussed in Chapter 2.

It is important to note that the use of a higher exposure limit based on the reference value for miners given as 2.4 Sv per J h $m^{-3}$ in Chapter 2 results in a much smaller region which is labelled adequate. Thus while this lower reference value may give a better estimate of the dose in the average mine than the ICRP value of 3.0 Sv per J h $m^{-3}$, its use in setting a higher exposure limit may require the routine monitoring of particle size and/or unattached fraction in some mines to insure that the dose is limited according to the criteria developed above.

The conversion factor between thoron daughter exposure and dose is less dependent on the unattached fraction and particle size, partly because these parameters are thought to vary over a smaller range, as is shown in Section 6 of Chapter 2. The effective dose equivalent per unit exposure derived in Chapter 2 for aerosols in the size range 0.3 µm - 0.2 µm is 0.6 - 1.2 Sv per J h $m^{-3}$; smaller sized aerosols having higher conversion factors. Since the range of this factor is within 1.5 times the ICRP reference value, it would appear that the measurement of exposure alone in most situations would be an adequate measure of dose provided the particle size is not less than 0.1 µm, assuming the ICRP conversion factor of 1.0 Sv per J h $m^{-3}$ is used to limit exposure.

General public exposure

The previous section gives a method of setting limits on the unattached fraction and the particle size for occupational exposures for which the exposure is an adequate measure of dose. Breathing rates and age were assumed to be fixed. In this section limits are given in relation to exposure of the general public, where a lower breathing rate is used to derive reference values and the age dependence given in Chapter 2 is also considered.

Table 3.1 approximates the median values given in Figure 2.28 for the conversion factor (Sv per J h $m^{-3}$) for adults exposed indoors to radon daughters as a function of breathing rate for low, intermediate and high ventilation rates, as defined in Section 2.5.1. Included in Table 3.1 are the breathing rates at which the conversion factor would exceed 1.5 times the recommended reference value of 1.5 Sv per J h $m^{-3}$ given in Table 2.11 for adult indoor exposure. It can be seen that except in dwellings with high ventilation rates (greater than one air change per hour) the conversion factor at a mean breathing rate of 0.75 $m^3$ $h^{-1}$ does not exceed 2.3 Sv per J h $m^{-3}$ and hence the measurement of exposure alone would be judged adequate.

Table 3.1

APPROXIMATE RELATIONSHIPS BETWEEN BREATHING RATE B ($m^3$ $h^{-1}$) AND THE FACTOR CONVERTING EXPOSURE TO DOSE (Sv PER J h $m^{-3}$), BASED ON DATA PRESENTED IN FIGURE 2.28.

The relationships are given for values of aerosol size and unattached fraction that are througt appropriate for conditions of low (< 0.3 $h^{-1}$), moderate (0.3 - 1.0 $h^{-1}$), and high (> 1 $h^{-1}$) rates of ventilation. The columns labelled $B_{lim}$ give breathing rates at which the conversion factor exceeds 2.3 Sv per J h $m^{-3}$

| Ventilation Class | J-E Model | | J-B Model | |
|---|---|---|---|---|
| | Sv per J h $m^{-3}$ | $B_{lim}$ | Sv per J h $m^{-3}$ | $B_{lim}$ |
| Low | 0.7 + 0.9 B | 1.8 | 0.2 + 1.5 B | 1.4 |
| Moderate | 0.7 + 1.4 B | 1.1 | 0.5 + 1.9 B | 1.0 |
| High | 0.7 + 2.4 B | 0.7 | 0.6 + 3.4 B | 0.5 |

Figure 2.31 summarized the results of calculations of the conversion factor for radon daughters as a function of age. Most of the values given exceeded 1.5 times the reference value of 1.5 Sv per J h $m^{-3}$ (Chapter 2). The reason for this is that the reference value assumed moderate to low ventilation, while Figure 2.31 was derived for moderate ventilation. It would seem reasonable, therefore, particularly since most children spend more time outdoors than adults, that if the measurement of exposure alone is adequate for adults it will be adequate for all age groups as the variations with age is not great. Also, if the reference value of 1.5 Sv per J h $m^{-3}$ is used, the dose will not be underestimated except where high ventilation rates persist, and exposure will in general be low.

Detailed calculations of conversion factors for thoron daughters as functions of breathing rates and ages are not given in Chapter 2. It is reasonable to assume that doses are directly proportional to breathing rates in adults (Section 2.6.4) and if the conversion factor 0.5 Sv per J h $m^{-3}$ is used for adult domestic exposure (c.f. the factor 1.0 Sv per J h $m^{-3}$ for occupational exposure), the doses will not be underestimated. Detailed age dependent calculations of the effective dose equivalent for thoron daughters is more difficult than for radon daughters as tissues other than the lung must be considered (see Table 2.13). Calculations using standard ICRP models (ICRP, 1979) extrapolated for use with a one year old infant (Johnson,1982) result in an effective dose equivalent for the infant about a factor of 10 above that for the adult per unit of inhaled Th-B activity. Since the average breathing rate in the one year old is almost a factor of 10 below that for an adult (Table 2.2) these factors will almost cancel out. It is reasonable to assume then that a value of 0.5 Sv per J h $m^{-3}$ will not underestimate the dose from indoor exposure to thoron daughters at any age.

### 3.2.2 Criteria for radiobiology and epidemiology

The above criteria developed to limit doses from exposures to radon and thoron daughters for radiation protection purposes are not necessarily applicable for the purposes of radiobiology and epidemiology. The reasons for this are several. For example, a bias may have been introduced in the process of insuring that the upper limit on dose has been sufficiently well estimated. Also, an individual investigator may wish to estimate average doses more accurately and the detailed dose distribution may be important in a particular study.

For these reasons it is not possible to develop overall criteria for radiobiological or epidemiological studies. Each investigator will have to decide individually, based on the information given in Chapter 2 or information developed elsewhere, if parameters other than the exposure alone need to be monitored. It is likely that a survey of such parameters as particle size and unattached fraction, and the selection of a suitable factor to convert from exposure to dose will suffice in most situations.

## 3.3 CONCLUSIONS

Based on the criteria developed above, the measurement of exposure and the application of two factors, one for occupational exposure and one for exposure of the general public, to convert this to dose will result in adequate radiation protection except for conditions which result in high unattached fractions for radon daughters and small particle sizes for thoron daughters.

The factor 3.0 Sv per J h $m^{-3}$ is adequate to convert radon daughter exposure to dose for occupational exposure unless particle size is significantly smaller or the unattached fraction is significantly larger than that normally encountered in mines. The factor 1.5 Sv per J h $m^{-3}$ is adequate for radiation protection of the general public at all ages, except in areas which have high ventilation rates (greater than one air change per hour) but also high levels of exposure.

A factor of 1.0 Sv per J h $m^{-3}$ is adequate for all occupational exposures to thoron daughters for radiation protection purposes. This factor may considerably overestimate the dose, but will not significantly underestimate it unless the particle size is smaller than 0.1 $\mu$m AMD. The factor 0.5 Sv per J h $m^{-3}$ is likewise adequate for radiological protection of the general public at any age.

Chapter 4

# REVIEW OF OBJECTIVES AND REQUIREMENTS FOR MEASUREMENT AND MONITORING

## 4.1 INTRODUCTION

The basis for any measures in radiation protection is to ensure that doses are below regulatory limits and, if below these limits, as low as reasonably achievable. Assessment of doses for radon and thoron daughters relies on measurement of the radiation exposure. For this assessment measurements of the environmental concentrations of relevant radionuclides are required. Due to the extensive requirements in health physics manpower and related costs, it is necessary to optimise measurement programs and to select the population groups to be monitored. The specific requirements for the measurement of atmospheric levels of radon, thoron and their daughters for practical radiation protection purposes differ significantly from those in the case of research applications, for example, assessment of exposure and dose in epidemiological studies.

In the following, objectives and requirements for the measurement of radon, thoron and their daughters are discussed for routine operational health physics and special research applications according to the scheme outlined in Figure 4.1.

## 4.2 OBJECTIVES OF MEASUREMENT AND MONITORING IN SPECIFIC ENVIRONMENTS

The main objectives of monitoring and measurement programmes in radiation protection are:

- to achieve and maintain safe and satisfactory working conditions for occupationally exposed groups

- to achieve and maintain safe environmental conditions for members of the general public

- compliance with national standards

- evaluation of the effectiveness of environmental control measures and equipment

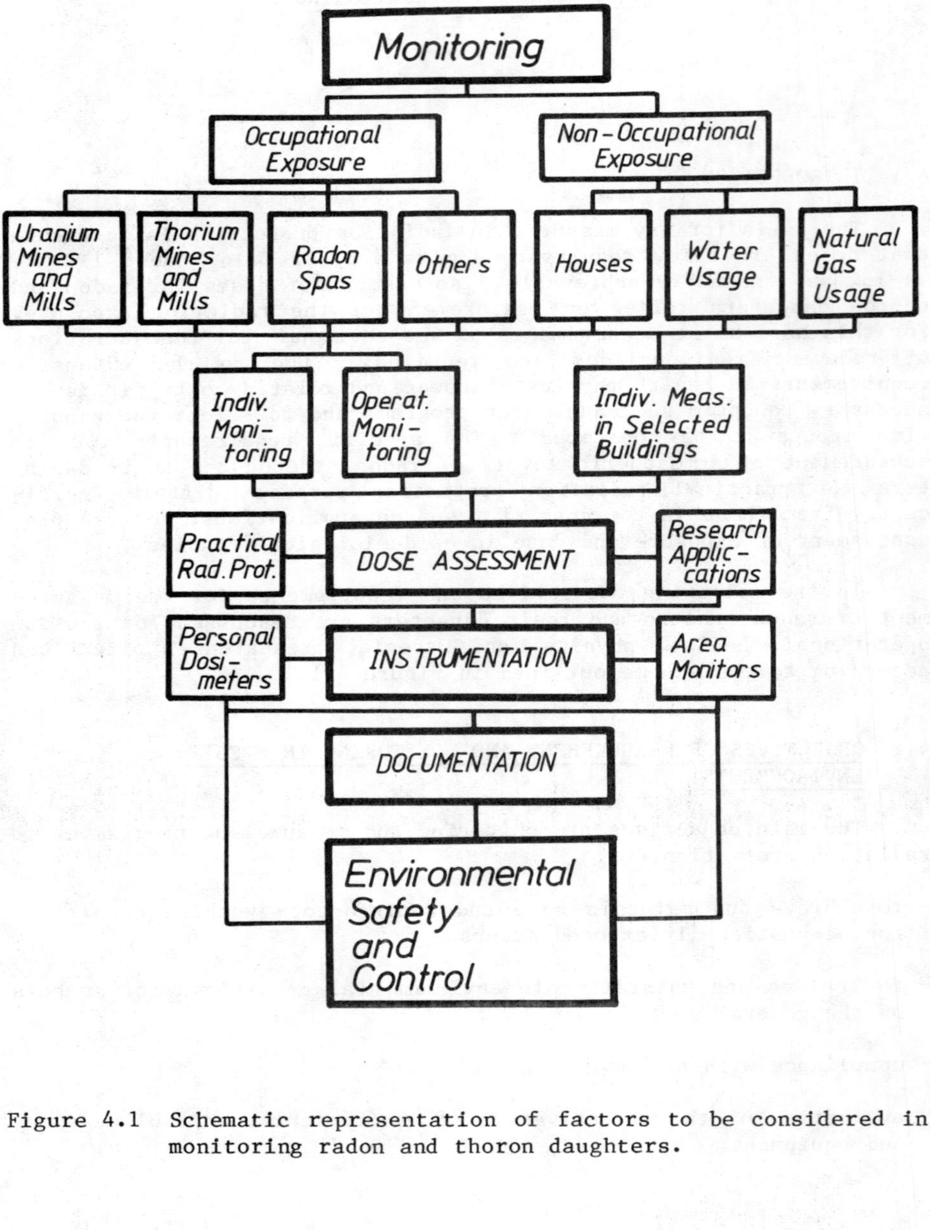


Figure 4.1 Schematic representation of factors to be considered in monitoring radon and thoron daughters.

- potential use of the information derived from such programmes in future scientific investigations

4.2.1 Environments

Based on recommendations by the International Atomic Energy Agency and International Labour Organisation, and results of international scientific meetings, the assessment of the concentration of radon, thoron and their daughters in air is required in the following situations (IAEA, 1976; ICRP, 1977; OECD, 1979; AECB, 1980; Clemente et al, 1981):

Occupational

uranium mining - inhalation hazard mostly in underground operations due to radon emanating from the ore body and, to a lesser extent, from radon contained in water dripping from roofs and walls, seepage at ground level, etc.

In the case of open-pit mining, inhalation hazards can arise with high grade ore bodies under conditions of temperature inversion, and the fraction of unattached radon daughters is likely to be high.

uranium milling - elevated levels can be found mainly near ore storage bins and crushing/grinding circuits

thorium mining - in underground mining increased exposure can occur from thoron decay products (mainly ThB) due to thoron emanating from the ore body

thorium milling - high concentrations may occur during initial chemical treatment, ore storage and crushing, drying of concentrate cake

thorium fabrication - high concentrations of thoron and daughters can arise in the manufacturing of gas mantles, thorium alloys and refractory crucibles.

radon spas - in radon spas either thermal spring water containing radon or water with artificially elevated radon content is used in treatment centres; undesirably high concentrations can arise for employees if insufficient ventilation causes radon build-up in the treatment areas

others - for underground non-uranium mines the situation is analogous to underground uranium operations, although the radon source may be difficult to identify; radon problems may also occur in tunnels or caves (eg, in hydro-electric power production or military installations) and in open-pit mines

Non-occupational

Construction material with elevated content of radium-226 and/or thorium-232 - manufacture and use of building materials with high natural radionuclide content or due to technological enhancement can result in increased concentrations of radon and thoron daughters indoors

dwellings with increased radon flux from the surrounding subsoil - soil with radioactive contamination from industrial operations, eg, from uranium ore processing (mill tailings), and with relatively high porosity can release radon gas. This can enter lived-in volumes through active transport or diffusion processes

energy-efficient dwellings - where low infiltration rates, decreased mechanical ventilation or passive solar heating (eg, bedrock thermal storage) are employed or in houses built underground in order to reduce energy consumption for heating and cooling, indoor air quality is generally lowered and radon levels as well as equilibrium ratios can be increased

radon-rich water - can contain a naturally increased concentration of radon, eg, in deep bored wells. Domestic and commercial use of the water deemanates radon into the indoor atmosphere

radon in domestic gas - natural gas can contain radon in the range of 0.04 to 2 $kBq/m^3$ (UNSCEAR, 1977). Burning natural gas for heating and cooking can result in measurable amounts of radon in homes.

### 4.2.2 Monitoring requirements

Taking the objectives of measurement and monitoring into consideration (Section 4.2) the following types of monitoring are required for the different environments mentioned above:

Occupational exposure

This type of exposure occurs mainly in mining and milling environments. It is characterised by large temporal and/or local fluctuations of the atmospheric radon and daughter concentrations, with even greater fluctuations for thoron and its daughters. Furthermore, time spent by the worker in a specific working area or at a particular manual operation can be of greatly varying duration. Therefore, individual monitoring of workers should be given preference over area monitoring if it can be shown that this leads to lower doses. High risk individuals who are assumed to receive higher exposures representative of a specific work routine should be monitored individually.

In addition operational monitoring of the work place is required to confirm appropriate control of the routine procedure or after operational changes and also to indicate the onset of abnormal conditions. This can be achieved with high cost-effectiveness by area monitoring at fixed locations. This type of monitoring is also useful in selected buildings in order to investigate high concentrations and to develop suitable remedial measures.

Non-occupational exposure

Members of the public can receive increased exposure to radon, thoron and their daughters predominantly in the indoor environment due to either natural or technologically enhanced sources. In order to identify areas with enhanced indoors levels of these radionuclides, large scale surveys are required. These can be carried out by individual measurements in selected dwellings considered representative for the housing of the population group studied. The criteria for selection can be: use of certain construction materials, ventilation and heating system, type of drinking water supply or geographical position of the building in a certain district with specific geological conditions of the subsoil. Furthermore, "life style" factors should be considered for the inhabitants, for example age dependent physical activity (Hofmann et al, 1980).

If the programme objective is to decide on compliance of the particular situation with adopted limits (regulations, clean-up criteria, etc), practical difficulties, like limitations of manpower, time and availability of financial resources impose a limitation of the amount of data to be collected. Since the nuclide concentrations also vary a great deal within a given structure, it is recommended to take individual measurements in a room where elevated levels are to be expected, ie, usually the basement, preferably with windows and doors closed. In this manner results will overestimate the true mean value, but the probability of erroneous decisions on compliance will be reduced.

## 4.3 DOSE ASSESSMENT

### 4.3.1 Practical radiation protection

For dose assessment with regard to practical radiation protection for workers, members of the public or specific population groups the determination of the potential alpha-energy intake or exposure is necessary in all cases; in the case of dwellings and radon spas and in most occupational situations it is also sufficient. In order to improve the accuracy of the dose assessment, it is adventageous to assess the respiratory minute volume and to measure in addition the activity median diameter of the aerosol and the free atom fraction of potential α-energy ($f_p$); the latter is of particular importance for open cut mines and milling operations.

### 4.3.2 Research applications

In addition to the above parameters, for research applications such as radiobiological interpretation of effects in animal experiments or epidemiological investigation of lung cancer incidence in humans, the following should also be taken into account:

a) biological variability amongst the exposed species (ratio nose/mouth breathing, target geometry in respiratory tract, clearance processes, age, sex)

b) nature of carrier aerosols and presence of known or suspected synergistic factors (condensation nuclei, diesel smoke, acid fumes, ore dust).

These variables can cause a regional and radionuclide-specific dose distribution in parts of the lung and thereby influence the conversion factor between exposure and dose (Chapter 2).

## 4.4 INSTRUMENTATION AND RECORD KEEPING

### 4.4.1 Personal dosimeters

Personal air sampling monitors should permit the assessment of the individual time-integrated exposure due to the alpha emitting radon and thoron daughters in the inhaled air (breathing zone monitoring). The instrument design must give reliable service even under severe mining conditions at economically feasible operating costs. At present there are only a few instruments available, which have been tested for their practical application, such as the CEA alpha dosimeter (Zettwoog, 1981), the US - electronic personal dosimeter (Durkin, 1979) and passive personal monitors (Domanski et al, 1982, Orzechowski et al, 1982).

### 4.4.2 Area monitors

The description of the average atmospheric concentration of radon, thoron and their daughters in a given area requires that samples are taken preferably continuously, or at least frequently. Currently either conventional filter sampling methods or fast WL-measurement techniques are used.

### 4.4.3 Record keeping

Monitoring records should provide the relevant information for the following tasks:

- assessment of the individual cumulative radon daughter exposure and resulting dose

- maintenance and/or improvement of present exposure conditions

- demonstration of compliance with regulations.

In order to calculate the individual occupational dose, it is suggested that standardized forms be used; for example, the modified Australian data form of Fig. 4.2 (Leach and Lokan, 1979). This would facilitate future use of data for scientific (epidemiological) studies or medico-legal matters (eg, litigation). The following data should be collected:

1. information about the employee (name, date of birth, employment number, personal dosimetric number and preferably personal respiratory minute volume as determined from annual medical examination simulating a specified work load).

NAME OF SITE: | NAME OF EMPLOYEE: | PERSONAL DOSIMETRIC RECORD NUMBER:

OPERATOR: | EMPLOYMENT NUMBER:

MINE MANAGER: | OCCUPATION: | PERIOD:

RESP. MINUTE VOLUME
REST: HEAVY WORK:

| DATE | WORK AREA | | POTENTIAL ALPHA ENERGY | | EFFECTIVE DOSE EQUIVALENT | ADDITIONAL ATMOSPHERIC PARAMETERS | | COMMENTS |
|---|---|---|---|---|---|---|---|---|
| | DESCRIPTION | TIME OCCUPIED (h) | AVERAGE CONCENTRATION ($J/m^3$) | EXPOSURE ($J\ h/m^3$) | (mSv) | $f_p$ (%) | AMD (μm) | |
| | | | | | | | | |

Figure 4.2 Example of individual dose record for occupational exposure to $^{222}Rn$-daughters.

2. information on occupation and estimated exposure (type of work, time spent at different work stations, J h $m^{-3}$, with AMD, $f_p$ at given stations for different times as applicable.)

3. data on location of employment and management, name of mine operator and manager.

Due to a latency period between 15 and 30 years for the induction of a radiation induced disease, records should be kept for a minimum of 30 years after the exposure related employment has been terminated.

### 4.4.4 Development need

In situations where the free atom fraction of potential α-energy, $f_p$, is high and variable there is a need for a personal dosimeter capable of measuring $f_p$, based for example on the work of George (1972).

## 4.5 CONCLUSIONS

Concern has been expressed about the hazard due to occupational and non-occupational exposure to atmospheric radon, thoron and their daughters. In order to minimize the potential health risk resulting from the inhalation of these nuclides, recommendations have been made (ICRP, 1981) concerning the upper limit of intake and exposure. The basis for practical implementation of and compliance with these limits is the measurement of representative samples for the given exposure conditions.

Objectives and requirements for practical radiation protection in industry differ from those for radiation protection of the public and for research orientated studies in a number of aspects, for example dosimetric requirements, sampling statistics and instrumentation. In occupational radiation protection two distinct objectives are identified; exposure assessment for individual workers to ensure compliance with exposure limits and area monitoring to maintain a safe working place and prevent high exposures. By defining monitoring programmes which take these differences into account cost-effectiveness can be increased and manpower requirements reduced.

It is recommended that an interdisciplinary approach is taken in the design of such measurement programmes, comprising of health physics, statistics and operational engineering. In this manner the data obtained will provide the information necessary for optimized radiation protection of both workers and the general public.

Chapter 5

# SUMMARY AND CONCLUSIONS

(1) Epidemiological surveys of uranium miners exposed in the past to high concentrations of radon daughters in air have shown that incidence of bronchogenic cancer is related to exposure to the potential $\alpha$-energy of the short-lived daughters. The epidemiological data, however, are not adequate as the sole basis for limiting risk from exposure to radon and thoron daughters in radiological protection (ICRP, 1981). An independent assessment of dose absorbed by sensitive tissues is required with summation of risks from all significant sources of tissue irradiation.

(2) Stem cells in bronchial epithelium are considered the cells at risk for induction of bronchogenic cancer, and therefore absorbed dose, averaged over all stem cells in the ciliated bronchial region of the lung, is the quantity determining probability of inducing bronchogenic cancer. For a given exposure to potential $\alpha$-energy, the dose absorbed by the bronchial region must depend on both the biological factors in the exposed individual and the physical conditions of exposure. This dependence can only be assessed by dosimetric modelling.

(3) The initial objective of this report was to review analytical modelling of dose absorbed by lung tissue, principally in the bronchial region, in relation to exposure to potential $\alpha$-energy. The review was based on the procedure developed by Jacobi and Eisfeld (1980) for modelling the average dose to stem cells throughout the bronchial and pulmonary regions of the lung and examining sensitivity of these calculated doses to biological parameters of the lung model. Their analysis has been extended by considering more recent data on bronchial dimensions in the human lung (Yeh and Schum, 1980); the depth distribution of bronchial stem cells and mucus thickness (Gastineau et al, 1972); and substantially different compartment models representing clearance of deposited radon and thoron daughters in mucus with retention of a fraction of daughter activity in epithelial tissue before uptake into the blood. The influence of age and breathing rate on regional lung dose calculated by models has also been examined.

(4) Different models of aerosol deposition, clearance and dosimetry have been applied to examine the dependence of regional lung dose on

the physical characteristics of radon and thoron daughter aerosols; the fraction of potential α-energy not attached to aerosol particles, $f_p$; the diffusion coefficient of these unattached atoms; the activity median diameter (AMD) of the attached daughter aerosol; and the radioactive equilibrium factor, F, between the daughters and the parent gas. Dose absorbed by the bronchial and pulmonary regions of the lung per unit exposure to potential α-energy was calculated to depend on $f_p$ and AMD, but to be independent of the equilibrium factor. In general terms regional lung dose is given by:

$$D_{T-B} = k_{T-B,a} (1-f_p) E_p + k_{T-B,f} f_p E_p$$

$$D_P = k_{P,a} (1-f_p) E_p$$

where the coefficients, k, represent dose per unit exposure for the attached fraction ($k_{T-B,a}$; $k_{P,a}$) and the free atom fraction ($k_{T-B,f}$) of the potential α-energy to which the individual is exposed.

The values of $k_{T-B,a}$ and $k_{P,a}$ depend on the AMD of the carrier aerosol; the biological factors breathing rate and age; and, to a lesser extent, the model used for calculation. The value of $k_{T-B,f}$ also depends on the parameters breathing rate and age, but in addition it is more sensitive to the assumptions made in modelling clearance of inhaled free atoms from the lung. In practice, however, different dosimetric models give similar estimates of bronchial dose, because the value of the unattached fraction of potential α-energy, $f_p$, expected in most exposure environments is low. Irrespective of the model employed for calculation, dose absorbed by the pulmonary region, $D_P$, is small compared with that absorbed by the bronchial region, $D_{T-B}$, in all cases of exposure to either radon or thoron daughters.

(5) To assess exposure in mines to radon daughters, a mean breathing rate of 1.2 $m^3 h^{-1}$ has been adopted. On the basis of the limited data available, a mean value of the unattached fraction, $f_p$, of about 0.02 (range 0.01-0.05) is expected for long-term exposure in working areas of underground mines. The expected range of activity median diameter (AMD) for the attached aerosol is 0.15-0.3 μm. Under these conditions the results of different dosimetric models are sufficiently similar to consider 6 mGy per WLM exposure to potential α-energy (1.8 Gy per J h $m^{-3}$) as a reference conversion factor between exposure and bronchial absorbed dose, with an uncertainty range of about ± 50%.

(6) A wider range of atmospheric conditions is expected in open-pit mines with occasional exposure to high unattached fractions of potential α-energy and small aerosol AMD. Under these more extreme conditions the conversion factor between exposure and bronchial absorbed dose may be as high as 15 mGy per WLM. In general and for long-term exposures, however, the reference conversion factor of 6 mGy per WLM may also be applicable for assessment of exposure to radon daughters in open-pit mines, provided that the fraction of

exposure due to unattached potential α-energy is not substantially greater than 0.05.

(7) To assess risk from intake of radon daughters for the purpose of occupational radiological protection, the ICRP has recommended (ICRP, 1981) that a quality factor of 20 be applied to the doses absorbed by the bronchial and pulmonary regions of the lung and the risk of fatal cancer estimated by applying a weighting factor of 0.06 to each tissue dose for summation of an effective dose equivalent, $H_E$. Using this procedure a reference value for effective dose equivalent of 8.5 mSv per WLM exposure (2.4 Sv per J h $m^{-3}$) is proposed to limit lifetime exposure to radon daughters in mines, both underground and open-pit, subject to the qualifications of paragraphs (13) through (17) below. A conversion factor of 8.5 mSv per WLM exposure is also expected to apply in non-uranium mines and in most other cases of occupational exposure to radon daughters where a mean breathing rate of about 1.2 $m^3$ $h^{-1}$ is appropriate.

(8) To assess domestic indoor exposure of adults to radon daughters a lower mean breathing rate of 0.75 $m^3$ $h^{-1}$ has been adopted. There are no reliable data on typical values of $f_p$ and aerosol AMD in domestic environments, and their dependence on room conditions. It is expected, however, that $f_p$ and AMD will vary with the rate of exchange between indoor and outdoor air but, in general, $f_p$ will be slightly higher and AMD smaller than in mine atmospheres. Under these conditions, a reference value of 4 mGy per WLM exposure (1.2 Gy per J h $m^{-3}$) is proposed for domestic exposure of adults. This value may be as high as 8 mGy per WLM in the case of a high exchange rate between indoor and outdoor air, but it is then most likely that exposure to radon daughters will be too low to be of concern. Under the assumed reference conditions, bronchial absorbed dose in children per unit exposure to radon daughters is calculated to be about a factor 1.5 higher than in a 30 year old adult. For older members of the public, however, bronchial absorbed dose will be correspondingly lower. Therefore it is proposed that a reference value of 4 mGy per WLM exposure can be used to assess exposure of the general public to radon daughters, without age correction. A reference value for effective dose equivalent of 5.5 mSv per WLM exposure (1.5 Sv per J h $m^{-3}$) is thus derived for members of the general public, using the weighting procedure recommended by the ICRP for assessing occupational exposure (ICRP, 1981).

(9) It is emphasised that dosimetric modelling can only provide estimates of dose absorbed by tissues and their sensitivity to values of the parameters incorporated in lung models. Although it is unlikely that exposure to co-carcinogenic or synergistic factors, particularly in mine atmospheres, will have a major effect on absorbed dose, it is almost certain that these factors do influence sensitivity of lung tissue to the induction and promotion of cancer. The numerical procedure for weighting absorbed dose under different environmental conditions in order to estimate risk is therefore a matter for judgement and must be justified by comparing these risk estimates with epidemiological evidence. By applying risk factors to absorbed dose consistent with the epidemiological evidence from

uranium miners, however, the estimated risk from inhalation of radon daughters in dwellings will most probably be conservatively high.

(10) In the case of both occupational and domestic exposure to thoron daughters, the aerosol characteristics are less well known, but aerosol AMD is probably larger than for radon daughters and $f_p$ lower. Under these conditions, for occupational exposure at a mean breathing rate of 1.2 $m^3 h^{-1}$, the conversion factor between exposure and mean bronchial absorbed dose is calculated to be in the range 1-2 mGy per WLM (0.3-0.6 Gy per J h $m^{-3}$). Essentially all of this dose arises from intake of thorium-B ($^{212}Pb$) attached to the ambient aerosol. Uncertainties in the biological parameters of lung models are more critical for thoron than radon daughters and are largely responsible for the quoted range in the dose conversion factor.

(11) Absorption of $^{212}Pb$ from the pulmonary region of the lung into the blood and uptake on bone surfaces and in the kidneys is calculated to contribute about equally with irradiation of the bronchial epithelium to an effective dose equivalent. Thus, the effective dose equivalent for occupational exposure to thoron daughters is expected to be in the range 2-4 mSv per WLM exposure (0.6-1.2 Sv per J h $m^{-3}$). A reference value of 3 mSv per WLM is proposed. This conversion factor is about one-third the value derived for occupational exposure to potential $\alpha$-energy from radon daughters. For exposure of adults to thoron daughters, the effective dose equivalent is approximately proportional to breathing rate. A reference conversion factor of 2 mSv per WLM exposure (0.5 Sv per J h $m^{-3}$) is therefore proposed for domestic exposure.

(12) Given adequate knowledge of the conditions typifying exposure in different environments, ie, aerosol size (AMD), unattached fraction of potential $\alpha$-energy ($f_p$), and breathing rate, the depth distribution of bronchial stem cells and the clearance characteristics of free radon daughter atoms are the most critical parameters influencing estimation of bronchial dose. By averaging dose over stem cells throughout the bronchial region, the importance of these parameters is minimised. However, the distribution of stem cell depths, the estimated dose contributed by free atoms and also that from the large particles in the aerosol deposited by their inertia become much more critical if it is required to specify dose absorbed in the hilar bronchi. It is in this part of the bronchial tree that bronchogenic cancers tend to be reported and exposure co-factors may also have their effect.

(13) Adequacy of the quantity 'exposure to potential $\alpha$-energy' as a measure of dose has been assessed by examining the variation in results of lung modelling with the additional measurable parameters characterising exposure. Thus, if it were shown that exposure to potential $\alpha$-energy alone gives an insufficiently accurate assessment of dose, monitoring of aerosol size distribution, free atom fraction or breathing rate may also be required. It was first necessary to develop criteria for the required accuracy and precision of internal dose assessment. Such criteria must take into account different requirements; firstly of radiological protection in limiting doses to give acceptable risk; and secondly of epidemiological studies,

to give an unbiased estimate of dose for comparison with observed risk. They must also take into account the precision and accuracy in the primary measurement or assessment of individual exposure; there being no benefit in requiring greater accuracy in the conversion between exposure and dose. Finally, the potential benefit of improving the dose assessment by requiring more complex monitoring must be balanced against monitoring costs and diversion of effort from the aim of reducing exposures.

(14) Based on these considerations, the criteria developed for adequacy of exposure measurement in occupational radiological protection was that the factors converting exposure to effective dose equivalent for the particular conditions of individual exposure should not be more than 50% greater than the conversion factors implied by the ICRP in recommending annual limits on exposure of 0.017 J h $m^{-3}$ and 0.05 J h $m^{-3}$ respectively, for the radon and thoron daughters (ICRP, 1981). Thus, monitoring and limitation of exposure will provide an adequate means of limiting annual dose if the conversion factor between exposure and effective dose equivalent does not exceed 4.5 Sv per J h $m^{-3}$ for the radon daughters and 1.5 Sv per J h $m^{-3}$ for the thoron daughters. This criterion is satisfied for occupational exposure to radon daughters, irrespective of the dosimetric model employed for the assessment, if the unattached fraction of potential $\alpha$-energy is less than about 0.05. The criterion is also satisfied for exposure to thoron daughters if the aerosol AMD is larger than about 0.1 $\mu$m. In almost all cases therefore, it is considered that monitoring of exposure to potential $\alpha$-energy and application of the limits recommended by the ICRP will ensure adequate occupational radiological protection.

(15) The exposure limit recommended by the ICRP for radon daughters implies a conversion factor of 3.0 Sv per J h $m^{-3}$. The dosimetric analysis in this report gave a reference factor of 2.4 Sv per J h $m^{-3}$ for exposure conditions thought to typify mines. This lower conversion factor could be used to justify a higher exposure limit. In this instance however, the range of unattached fraction and aerosol size satisfying the criterion for adequacy of exposure as an estimator of dose is substantially restricted and it would be necessary to monitor these additional parameters in some cases.

(16) If there is evidence in open-pit mining that the unattached fraction of potential $\alpha$-energy typically exceeds 0.05 and also that annual exposures are likely to approach the derived limit (ICRP, 1981 and paragraph (14)), the aerosol characteristics, AMD and $f_p$, should also be measured over a suitable period of time. A more appropriate conversion factor between exposure and effective dose equivalent should then be derived using the data of figure 3.1 and the criterion for 'adequacy' of exposure measurement (paragraph (13)), in order to limit exposure in compliance with the basic dose limit and to the required accuracy of dose assessment. It may be appropriate for this purpose to adopt the results of the more conservative dosimetric model illustrated in figure 3.1 in deriving a conversion factor.

(17) In underground mining, where local control measures such as air cleaning by electrostatic precipitation or dilution by fresh atmospheric air are deemed necessary to reduce exposures to potential $\alpha$-energy, changes in the size distribution of the radon daughter aerosol, particularly an increase in $f_p$ and consequent increase in the factor converting exposure to dose, should be evaluated and considered as part of the optimisation process. For this application it may be appropriate to derive a conversion factor between exposure and effective dose equivalent from the mean of the results given in figure 3.1 for different dosimetric models.

(18) For radiological protection of the general public against radon daughter exposure a reference conversion factor of 1.5 Sv per J h $m^{-3}$ was proposed. In this case measurement of exposure alone is judged adequate, except in the rare circumstances of dwellings with high ventilation rates and also high exposures to radon daughters. For population exposure to thoron daughters, measurement of exposure alone is again judged adequate, with application of a conversion factor of 0.5 Sv per J h $m^{-3}$. Confirmation by reliable data of the size distribution of radon and thoron daughter aerosols and also the unattached fraction for radon daughters, typical of indoor atmospheres, is necessary for a more confident assessment of population dose. Likewise data on indoor occupancy and breathing rates as a function of age are required to improve the assessment of exposure to and intake of potential $\alpha$-energy.

(19) For the purpose of interpreting epidemiological data, unbiased estimates of dose should be used. These may require investigation of typical unattached fractions of potential $\alpha$-energy, aerosol size, exposure co-factors and breathing rates, if this can be justified by the precision of the exposure data and risk estimate for bronchogenic cancer obtained from the study population.

(20) The objectives of measurement and monitoring must be to achieve and maintain safe working conditions for occupationally exposed groups and safe environmental conditions for members of the general public. Monitoring must also demonstrate compliance with national dose standards. It should enable the effectiveness of environmental control measures to be evaluated and also provide estimates of exposure suitable for future analysis of epidemiological data. Considering these objectives, monitoring for occupational exposure should combine operational monitoring for control of the workplace with individual monitoring of high risk workers. For non-occupational exposure it is first necessary to identify areas or types of housing with enhanced levels of radon and thoron daughter exposure, for which individual measurements in selected dwellings are required. Subsequent monitoring must then be concerned with achieving or demonstrating compliance with national standards.

(21) The basis for assessment of dose and for practical implementation of dose limits is the measurement of representative samples for the individual exposure conditions. For practical radiation protection of workers or members of the public determination of the potential $\alpha$-energy exposure is necessary in all cases, and for most occupational exposure situations and in dwellings it is also

sufficient. The balance of effort between workplace monitoring and measurement of individual exposure, preferably by personal devices, must be optimised within the constraints of costs and resources, bearing in mind the levels of exposure likely to be incurred.

(22) Monitoring records should provide the information needed to assess individual cumulative exposure to potential α-energy from radon and thoron daughters and the resulting effective dose equivalent, including any contribution from external γ-irradiation and a notification or estimate of exposure to other known co-carcinogenic factors. Records should be organised so that derived exposure standards can be maintained and, if necessary, improved and compliance with regulations demonstrated. It is recommended that records of exposure to potential α-energy and other relevant or qualifying information should be kept in a standard form.

## Appendix A

## SPECIAL QUANTITIES AND UNITS

### (1) Radioactive decay properties

Radon gas ($^{222}Rn$) is formed by the decay of $^{226}Ra$ ($T_r$ = 1600 y) in the $^{238}U$-decay series. Thoron gas ($^{220}Rn$) is a radionuclide in the $^{232}Th$-decay series and is formed by decay of $^{224}Ra$ ($T_r$ = 3.64 d). Radon has a radioactive half-life ($T_r$) of 3.823 d and thoron 55 s. The main radioactive properties of the chains of nuclides resulting from decay of radon and thoron are given in Tables A.1 and A.2, respectively. The radioactive half-life ($T_r$), decay constant ($\lambda_r$), the principal energy emissions and their relative intensities are here reproduced from the data of UNSCEAR (1977).

Table A.1

MAIN RADIOACTIVE DECAY PROPERTIES OF $^{222}Rn$ AND ITS SHORT-LIVED DAUGHTERS (FROM UN 77)[1]

| Radionuclide decay chain | $T_r$ | $\lambda_r$ ($h^{-1}$) | Main energies (in MeV) and intensities: α | β | γ |
|---|---|---|---|---|---|
| $^{222}Rn$(Rn) ↓ α | 3.823 d | 7.55 $10^{-3}$ | 5.49 (100%) | - | 0.51 (0.07%) |
| $^{218}Po$(RaA) ↓ α | 3.05 min | 13.6 | 6.00 (~100%) | 0.33 (~0.019%) | - |
| $^{214}Pb$(RaB) ↓ β,γ | 26.8 min | 1.55 | - | 0.65 (50%)<br>0.71 (40%)<br>0.98 (6%) | 0.295 (19%)<br>0.352 (36%) |
| $^{214}Bi$(RaC) ↓ β,γ | 19.7 min | 2.11 | 5.45 (0.012%)<br>5.51 (0.008%) | 1.0 (23%)<br>1.51 (40%)<br>3.26 (19%) | 0.609 (47%)<br>1.12 (17%)<br>1.76 (17%) |
| $^{214}Po$(RaC') ↓ α | 164 μs | 1.52 $10^{7}$ | 7.69 (100%) | - | 0.799 (0.014%) |

[1]The branchings from $^{218}Po$ and $^{214}Po$ can be neglected, due to their low yield of 0.02%

Table A.2

MAIN RADIOACTIVE DECAY PROPERTIES OF $^{220}Rn$ AND ITS DAUGHTERS (FROM UN 77)

| Radionuclide decay chain | $T_r$ | $\lambda_r$ ($h^{-1}$) | Main energies (in MeV) and intensities | | |
|---|---|---|---|---|---|
| | | | α | β | γ |
| $^{220}Rn$(Tn) | 55 s | 45.4 | 6.29 (100%) | - | 0.55 (0.07%) |
| ↓ α | | | | | |
| $^{216}Po$(ThA) | 0.15 s | $1.58\ 10^4$ | 6.78 (100%) | - | - |
| ↓ α | | | | | |
| $^{212}Pb$(ThB) | 10.64 h | 0.06514 | - | 0.346 (81%)<br>0.586 (14%) | 0.239 (47%)<br>0.300 (3.2%) |
| ↓ β,γ | | | | | |
| $^{212}Bi$(ThC)<br>64% β,γ ; 36% α | 60.6 min | 0.686 | 6.05 (25%)<br>6.09 (10%) | 1.55 (5%)<br>2.26 (55%) | 0.040 (2%)<br>0.727 (7%) |
| $^{212}Po$(ThC') → α | 304 ns | $8.21\ 10^9$ | 8.78 (100%) | - | - |
| $^{208}Tl$(ThC") → β,γ | 3.10 min | 13.4 | - | 1.28 (25%)<br>1.52 (21%)<br>1.80 (50%) | 0.511 (23%)<br>0.583 (86%)<br>0.860 (12%)<br>2.614 (100%) |

(2) Potential α-energy

The potential α-energy of an atom, $\varepsilon_p$, is the total α-energy emitted during decay of this atom through the decay chain to $^{210}Pb$ (RaD) or $^{208}Pb$ (ThD), respectively. The total potential α-energy per Bq of activity of a radionuclide is given by $\varepsilon_p/\lambda_r$, where the decay constant $\lambda_r$ is expressed in $s^{-1}$. Values of $\varepsilon_p$ and $\varepsilon_p/\lambda_r$ for the radon and thoron decay chains are given in Table A.3.

Table A.3

POTENTIAL $\alpha$-ENERGY PER ATOM AND PER Bq

| Radionuclide | Potential $\alpha$-energy | | | |
|---|---|---|---|---|
| | per atom ($\varepsilon_p$) | | per Bq ($\varepsilon_p/\lambda_r$) | |
| | in MeV | in $10^{-12}$ J | in MeV | in $10^{-10}$ J |
| $^{222}$Rn(Rn) | 19.2 | 3.07 | 9.15 $10^6$ | 14 700 |
| $^{218}$Po(RaA) | 13.7 | 2.19 | 3 620 | 5.79 |
| $^{214}$Pb(RaB) | 7.69 | 1.23 | 17 800 | 28.6 |
| $^{214}$Bi(RaC) | 7.69 | 1.23 | 13 100 | 21.0 |
| $^{214}$Po(RaC') | 7.69 | 1.23 | 2.0 $10^{-3}$ | 3.0 $10^{-6}$ |
| $^{220}$Rn(Tn) | 20.9 | 3.34 | 1 660 | 2.65 |
| $^{216}$Po(ThA) | 14.6 | 2.34 | 3.32 | 5.32 $10^{-3}$ |
| $^{212}$Pb(ThB) | 7.80 | 1.25 | 4.31 $10^5$ | 691 |
| $^{212}$Bi(ThC) | 7.80 | 1.25 | 4.09 $10^4$ | 65.6 |
| $^{212}$Po(ThC') | 8.78 | 1.41 | 3.85 $10^{-6}$ | 6.2 $10^{-9}$ |

(3) Potential $\alpha$-energy concentration in air

The potential $\alpha$-energy concentration of any mixture of short-lived $^{222}$Rn- or $^{220}$Rn-daughters is the sum of the potential $\alpha$-energy of all daughter atoms present per unit volume of air. This quantity can be expressed in SI-units:

$$1 \text{ J m}^{-3} = 6.24 \times 10^{12} \text{ MeV m}^{-3} = 6.24 \times 10^{9} \text{ MeV l}^{-1}$$

The special unit 1 WL (Working Level) is often used for this quantity:

$$1 \text{ WL} = 1.3 \times 10^{5} \text{ MeV l}^{-1} = 2.08 \times 10^{-5} \text{ J m}^{-3}$$

1 WL corresponds approximately to the potential $\alpha$-energy concentration of short-lived $^{222}$Rn-daughters in air which are in radioactive equilibrium with a $^{222}$Rn-activity concentration of 100 pCi $l^{-1}$ = 3.7 Bq $l^{-1}$ = 3700 Bq $m^{-3}$.

For $^{220}$Rn-daughters in radioactive equilibrium with $^{220}$Rn, 1 WL corresponds to a $^{220}$Rn-concentration of 7.43 pCi $l^{-1}$ = 275 Bq $m^{-3}$.

In Table A.4 the conversion factors between activity concentration (in Bq $m^{-3}$) and potential $\alpha$-energy concentration are listed for the short-lived daughter nuclides of $^{222}$Rn and $^{220}$Rn.

Table A.4

POTENTIAL α-ENERGY CONCENTRATION PER Bq $m^{-3}$

| Radionuclide | MeV $l^{-1}$ | $10^{-10}$ J $m^{-3}$ | $10^{-6}$ WL |
|---|---|---|---|
| $^{218}$Po(RaA) | 3.62 | 5.79 | 27.8 |
| $^{214}$Pb(RaB) | 17.8 | 28.6 | 137 |
| $^{214}$Bi(RaC) | 13.1 | 21.0 | 101 |
| $^{214}$Po(RaC') | 2.0 $10^{-6}$ | 3.0 $10^{-6}$ | 1.6 $10^{-5}$ |
| $^{216}$Po(ThA) | 3.32 $10^{-3}$ | 5.32 $10^{-3}$ | 0.0256 |
| $^{212}$Pb(ThB) | 431 | 691 | 3 320 |
| $^{212}$Bi(ThC) | 40.9 | 65.6 | 315 |
| $^{212}$Po(ThC') | 3.85 $10^{-9}$ | 6.2 $10^{-9}$ | 3.0 $10^{-8}$ |

(4) Equilibrium equivalent radon concentration ($EC_{Rn}$) and equilibrium factor (F)

The $EC_{Rn}$ of a non-equilibrium mixture of short-lived Rn-daughters in air is that activity concentration of Rn in radioactive equilibrium with its short-lived daughters that has the same potential α-energy concentration, $c_p$. Therefore for $^{222}$Rn and its daughters:

$$EC_{Rn}\ (\text{Bq m}^{-3}) = 2.85 \times 10^{-2}\ c_p\ (\text{MeV l}^{-1})$$

$$= 1.78 \times 10^{8}\ c_p\ (\text{J m}^{-3})$$

$$= 3700\ c_p\ (\text{WL})$$

[ie, $EC_{Rn}$ (pCi $l^{-1}$) = 100 $c_p$ (WL)]

and for $^{220}$Rn and its daughters:

$$EC_{Tn}\ (\text{Bq m}^{-3}) = 2.12 \times 10^{-3}\ c_p\ (\text{MeV l}^{-1})$$

$$= 1.32 \times 10^{7}\ c_p\ (\text{J m}^{-3})$$

$$= 275\ c_p\ (\text{WL})$$

The "equilibrium factor", F, with respect to potential α-energy is defined as the ratio of the $EC_{Rn}$ to the actual activity concentration, $c_{Rn}$, of radon in air:

$$F = \frac{EC_{Rn}}{c_{Rn}}$$

(5) Activity and potential α-energy exposure (E)

The "activity exposure" of an individual to $^{222}$Rn or $^{220}$Rn is the time-integral over the activity concentration of $^{222}$Rn or $^{220}$Rn, respectively, to which the individual is exposed during a definite period of time. Its unit is for example Bq h $m^{-3}$.

The equivalent quantity for the short-lived $^{222}Rn$- or $^{220}Rn$-daughters is the "potential α-energy exposure" of an individual during a definite period of time. This quantity can be expressed in the units

$$1 \text{ J h m}^{-3} = 6.24 \times 10^{9} \text{ MeV h l}^{-1} = 4.80 \times 10^{4} \text{ WL h}$$

$$1 \text{ WL h} = 1.3 \times 10^{5} \text{ MeV h l}^{-1} = 2.08 \times 10^{-5} \text{ J h m}^{-3}$$

The potential α-energy exposure of miners is often expressed in the unit 1 WLM (Working Level Month). 1 WLM corresponds to an exposure of 1 WL during the reference working period of 1 month (2000 working hours per year/12 months ≈ 170 h):

$$1 \text{ WLM} = 170 \text{ WL h} = 2.2 \times 10^{7} \text{ MeV h l}^{-1} = 3.5 \times 10^{-3} \text{ J h m}^{-3}$$

$$1 \text{ J h m}^{-3} = 285 \text{ WLM}$$

A potential α-energy exposure of 1 WLM $^{222}Rn$-daughters is associated with an $EC_{Rn}$ activity exposure of $6.3 \times 10^{5}$ Bq h $m^{-3}$.

(6) Activity and potential α-energy intake by inhalation (I)

The "potential α-energy intake" of an individual by inhalation of radon-daughters is the inhaled potential α-energy of the daughter mixture during a definite period of time. If $\bar{B}$ is the mean breathing rate during this period, the potential α-energy intake, $I_p$, is related to the potential α-energy exposure, $E_p$, by:

$$I_p = \bar{B} . E_p$$

For the reference worker ($\bar{B} = 1.2\ m^3\ h^{-1}$) the following conversion factors apply:

| Exposure, $E_p$ | → Intake, $I_p$ |
|---|---|
| 1 MeV h $m^{-3}$ | corresponds to $1.92 \times 10^{-13}$ J |
| 1 WL h | corresponds to $2.5 \times 10^{-5}$ J |
| 1 WLM | corresponds to $4.2 \times 10^{-3}$ J |
| 1 J h $m^{-3}$ | corresponds to 1.2 J |

For the reference adult exposed in a house ($\bar{B} = 0.75\ m^3\ h^{-1}$) the following conversion factors apply:

| Exposure, $E_p$ | → Intake, $I_p$ |
|---|---|
| 1 MeV h $m^{-3}$ | corresponds to $1.2 \times 10^{-13}$ J |
| 1 WL h | corresponds to $1.56 \times 10^{-5}$ J |
| 1 WLM | corresponds to $2.63 \times 10^{-3}$ J |
| 1 J h $m^{-3}$ | corresponds to 0.75 J |

"Activity intake" by inhalation is the inhaled activity of a radionuclide during a definite period of time. Intake of potential $\alpha$-energy, $I_p$, is related to intake of activity, $I_a$, by:

$$I_p = \left( \frac{E_p}{\lambda_r} \right) . I_a$$

The conversion factor $E_p/\lambda_r$ is the potential $\alpha$-energy per unit of activity of the daughter nuclide being considered. The following rounded values of the ratio $(I_p/I_a)$ and $(I_a/I_p)$ are recommended for practical purposes (ICRP, 1981).

Table A.5

CONVERSION FACTORS BETWEEN ACTIVITY INTAKE ($I_a$) AND POTENTIAL $\alpha$-ENERGY INTAKE ($I_p$) FOR $^{222}$Rn- AND $^{220}$Rn-DAUGHTERS

| Radionuclide | $\frac{I_p (10^{-10}\ J)}{I_a (Bq)}$ | $\frac{I_a (10^{8}\ Bq)}{I_p (J)}$ |
|---|---|---|
| $^{218}$Po(RaA) | 5.8 | 17.2 |
| $^{214}$Pb(RaB) | 28.6 | 3.50 |
| $^{214}$Bi(RaC) | 21.0 | 4.76 |
| $^{216}$Po(ThA) | 0.0053 | 18 900 |
| $^{212}$Pb(ThB) | 691 | 0.145 |
| $^{212}$Bi(ThC) | 65.6 | 1.52 |

## Appendix B

## DOSE FROM INHALED RADON, THORON AND DAUGHTERS TRANSLOCATED TO BODY ORGANS

Dosimetry of lung tissue for inhaled radon and thoron gas and dosimetry of other organs to which the gases are carried dissolved in blood and to which Rn-daughters deposited in lung can also be translocated has been reviewed by Jacobi and Eisfeld (1980). Their results are summarised in this appendix.

### (1) Inhaled $^{222}Rn$ and $^{220}Rn$

In order to calculate dose rates to tissues in equilibrium with a given concentration of $^{222}Rn$ or $^{220}Rn$ in inhaled air, it is assumed that:

(i) The short-lived $^{218}Po$-atoms formed by decay of Rn-atoms in lung air are all deposited on the alveolar epithelium where they decay.

(ii) $^{222}Rn$ dissolved in lung tissue is in radioactive equilibrium with all of its short-lived daughters. Dissolved $^{220}Rn$ however is in equilibrium only with $^{216}Po$, since the long half-life of $^{212}Pb$ enables this nuclide to be translocated to other tissues via the blood.

(iii) Daughter nuclides resulting from decay of $^{222}Rn$ and $^{220}Rn$ in the blood are partly transferred to solid tissues according to the uptake and retention parameters recommended by the ICRP (ICRP, 1979).

(iv) The short-lived daughters resulting from decay of $^{222}Rn$ dissolved in solid tissues attain radioactive equilibrium in these tissues.

(v) The volumetric saturation ratio of the $^{222}Rn$-concentration in all radiosensitive tissues relative to air is 0.4. Because of its short half-life, uptake of $^{220}Rn$ in all tissues except the lungs is negligible.

The resulting values for dose to tissues for exposure to the equilibrium Rn- concentration in air ($EC_{Rn}$) equivalent to a potential α-energy exposure of 1 J h $m^{-3}$ are given in Table B.1.

Table B.1

ESTIMATED DOSE TO TISSUE FOR EXPOSURE TO THE EQUILIBRIUM CONCENTRATIONS ($EC_{Rn}$) OF $^{222}Rn$ AND $^{220}Rn$ IN INHALED AIR EQUIVALENT TO 1 J h $m^{-3}$ POTENTIAL α-ENERGY

| Tissue | Dose/$EC_{Rn}$ in mGy per J h $m^{-3}$ | |
|---|---|---|
| | $^{222}Rn$ | $^{220}Rn$ |
| Lungs | 6.8 | 0.4 |
| Liver | 0.9 | 0.04 |
| Kidneys | 1.0 | 0.15 |
| Spleen | 0.9 | - |
| Red bone marrow | 0.8 | 0.02 |
| Bone surfaces | 0.8 | 0.3 |
| Body fat | 11 | - |
| Other tissues | 0.8 | - |

In the case of inhaled $^{222}Rn$, the lung and body fat are estimated to absorb the highest doses per unit exposure, but the dose to body fat can be neglected because this tissue is not radiosensitive. The lung is estimated to absorb about 7 mGy from exposure to the equilibrium concentration of radon equivalent to 1 J h $m^{-3}$ potential α-energy. Lung dose would be as high as 70 mGy per J h $m^{-3}$ exposure at an equilibrium factor, F, of 0.1, which can be regarded as a practical limit. Even this dose, however, can be neglected in comparison with dose to bronchial tissue from the inhaled $^{222}Rn$-daughters. This is estimated to be greater than 1 Gy per J h $m^{-3}$ (Chapter 5).

For the $^{220}Rn$-daughters, the equilibrium factor is about 0.02. Hence lung tissue is estimated to absorb about 20 mGy from inhaled $^{220}Rn$ gas per J h $m^{-3}$ exposure to the daughters, with somewhat lower doses to bone surfaces and the kidneys (Table B.1). Again, however, these tissue doses from the inhaled gas can be neglected in comparison with the dose to bronchial epithelium from the inhaled daughters, which is estimated to be higher than 300 mGy per J h $m^{-3}$ exposure (Chapter 5).

(2) Inhaled $^{222}Rn$- and $^{220}Rn$-daughters

Because of the higher regional deposition, uptake of $^{222}Rn$- and $^{220}Rn$-daughters from the lung to blood occurs predominantly in the pulmonary region. The fractions of deposited Po, Pb and Bi atoms taken up by the blood are given by the competing rates of radioactive decay and biological transfer represented by compartment models (Chapter 2, Section 2.2.4). The clearance models considered in the dosimetric review assume similar rates of biological transfer from the pulmonary region to the blood. On this basis, therefore, the fraction of inhaled Po, Pb and Bi atoms taken up by the blood depends mainly on the estimate of pulmonary deposition.

According to the recommendations of ICRP the daughter nuclides are retained in circulating blood with the following biological half-times (ICRP, 1979):

| Element (Nuclide) | $T_{bl}$ (d) |
|---|---|
| Po (RaA, ThA) | 0.25 |
| Pb (RaB, ThB) | 0.25 |
| Bi (RaC, ThC) | 0.01 |

Uptake of the daughter atoms in tissues is estimated from the competing rates of radioactive decay and biological elimination from the blood, assuming the following fractional uptakes for each element (Jacobi and Eisfeld, 1980):

Table B.2

UPTAKE FACTORS, f, OF TISSUES FOR Po, Pb and Bi ASSUMED BY JACOBI AND EISFELD (1980) BASED ON RECOMMENDATIONS OF ICRP (1979)

| Tissue | Element | | |
|---|---|---|---|
| | Po | Pb | Bi |
| Bone | 0 | 0.55 | 0 |
| Liver | 0.1 | 0.25 | 0.05 |
| Kidneys | 0.1 | 0.02 | 0.4 |
| Spleen | 0.1 | <0.01 | 0.01 |
| Rest of Body | 0.7 | 0.18 | 0.54 |

The uptake factor given for the rest of the body in Table B.2 represents activity distributed uniformly over all other soft tissues.

It is assumed that the activity in blood is uniformly distributed over the total mass m = 65 kg of soft tissue in the reference man (ICRP, 1975), including bone marrow.

The total activity in a tissue for determining dose is therefore given by the activity taken up from blood plus the contribution of blood activity to this tissue.

For inhalation of $^{222}$Rn-daughters Jacobi and Eisfeld (1980) have shown that the dose to all tissues except the kidneys is more than two orders of magnitude lower than that absorbed by lung tissue. The kidney dose can also be neglected, however, as this is estimated to be only a few percent of lung dose.

In the case of inhaled $^{220}$Rn-daughters, dose to bone surfaces and the kidneys from deposited $^{212}$Pb is estimated to be comparable with that absorbed by the bronchial region. The weighted dose equivalent contributed by this irradiation must therefore be included with that from lung irradiation to derive an effective dose equivalent. Dose to all other tissues and from deposited $^{212}Bi$ can be neglected.

## REFERENCES

AECB, 1980, Proceedings Third Workshop on Radon and Radon Daughters in Urban Communities Associated with Uranium Mining and Processing, Atomic Energy Control Board of Canada, Port Hope, Canada.

Albert, R.E., Lippmann, M. and Peterson, H.T., 1971, "The Effects of Cigarette Smoking on the Kinetics of Bronchial Clearance in Humans and Donkeys", Inhaled Particles III (Ed W. H. Walton), Pergamon, Oxford, pp 165-181.

Altshuler, B., Nelson, N and Kuschner, M., 1964, "Estimation of lung tissue dose from the inhalation of radon and daughters". Health Phys. 10, 1137-1161.

Archer, V.E., 1978, "Summary of data on uranium miners", In Workshop on Dosimetry for Radon and Radon Daughters, Oak Ridge National Laboratory, April 12-13, 1977. ORNL-5348 pp 23-25.

Armstrong, T.W. and Chandler, K.C., 1973, "SPAR, a FORTRAN Program for Computing Stopping Powers and Ranges for Muons, Charged Pions, Protons and Heavy Ions," ORNL-4869.

Bianco, A., Gibb, F.R. and Morrow, P.E., 1974, "Inhalation Study of a Submicron Size Lead-212 Aerosol," USAEC CONF-730907, pp 1214-1219.

Booker, D.V., Chamberlain, A.C., Newton, D. and Stott, A.N.B., 1969, "Uptake of Radioactive Lead following Inhalation and Injection," Br. J. Radiol. 42, 457-466.

Busigin, A., van der Vooren, A.W., Babcock, J.C. and Phillips, C.R., 1981, "The Nature of Unattached RaA ($^{218}$Po) Particles," Health Phys. 40, 333-344.

Chamberlain, A.C. and Dyson, E.D., 1956, "The Dose to the Trachea and Bronchi from the Decay Products of Radon and Thoron," Br. J. Radiol. 29, 317-325.

Chamberlain, A.C., Heard, M.J., Little, P. et al, 1978, "Investigations into Lead from Motor Vehicles," AERE-9198.

Cihak, R.W., Ishimaru, T. and Steer, A. et al, 1974, "Autopsy Findings and Relation to Radiation," Cancer, 33, 1580-1588.

Clemente, G., Steinhausler, F and Wrenn, M.E. (Eds), 1982, Proc. Specialist Meeting. The Assessment of Radon and Daughter Exposure and Related Biological Effects, Rome, Italy. R D Press, Univ. of Utah.

Crawford, D.J., 1981, "A generalized age-dependent lung model with applications to radiation standards". ORNL/NUREG/TM-411.

Davies, C.N., 1972, "Breathing of Half-micron Aerosols II Interpretation of Experimental Results," J. Appl. Physiol. 32, 601-611.

Davies, C.N., 1974, "Size Distribution of Atmospheric Particles", Aerosol Sci. 5, 293-300.

Domanski, T., Swiatnicki, G and Chruscielewski, W., 1982, "Comparison of the precision of measurement of radon and radon daughter exposure using simple passive track detectors and passive differentiating detectors". "Radiation Hazards in Mining: Control, Measurement, and Medical Aspects" Ed M. Gomez. Soc. of Mining Engineers, New York, NY, pp 409-418.

Duggan, M.J. and Howell, D.M., 1969, "Relationship between the Unattached Fraction of Airborne RaA and the Concentration of Condensation Nuclei," Nature, 224, 1190-1191.

Durkin, J., 1979, "Electronic radon daughter dosimetry". Health Phys. 37, 757-764.

Foster, W.M., Langenback, E.G. and Bergofsky, E.H., 1981, "Lung Mucociliary Function in Man: Interdependence of Bronchial and Tracheal Mucus Transport Velocities with Lung Clearance in Bronchial Asthma and Healthy Subjects," Inhaled Particles V, (Ed. W.H. Walton), Pergamon, Oxford, pp 227-244.

Fry, R.M., 1977, "Radon and its Hazards," in Personal Dosimetry and Area Monitoring Suitable for Radon and Daughter Products," Proc. NEA Specialist Meeting, Elliot Lake, Canada, 4-8 October, 1976, OECD-NEA, pp 13-32.

Gastineau, R.M., Walsh, P.J. and Underwood, N., 1972, "Thickness of Bronchial Epithelium with Relation to Exposure to Radon," Health Phys. 23, 857-860.

George, A.C., 1972, "Measurements of uncombined radon daughters in uranium mines". Health Phys. 23, 791-803.

George, A.C. and Breslin, A.J., 1969, "Deposition of Radon Daughters in Humans Exposed to Uranium Mine Atmospheres," Health Phys. 17, 115-124.

George, A.C., Hinchcliffe, L. and Sladowski, R., 1975, "Size Distribution of Radon Daughter Particles in Uranium Mine Atmospheres," Am. Ind. Assoc. J., 34, 484-490.

George, A.C., Hinchcliffe, L. and Sladowski, R., 1977, "Size Distribution of Radon Daughter Particles in Uranium Mine Atmospheres," Rep. HASL-326.

Gormley, P.G. and Kennedy, M., 1949, "Diffusion from a Stream Flowing Through a Cylindrical Tube," Proc. Roy. Irish Acad. 52, 163.

Greenhalgh, J.R., Birchall, A., James, A.C. et al, 1982, "Differential Retention of $^{212}Pb$ Ions and Insoluble Particles in Nasal Mucosa of the Rat". Phys. Med. Biol. 27, 837-851.

Haque, A.K.M.M and Collinson, A.J.L., 1967, "Radiation dose to the respiratory system due to radon and its daughter products". Health Phys. 13, 431-443.

Harley, N.H. and Pasternack, B.S., 1972, "Experimental Absorption Applied to Lung Dose from Thoron Daughters," Health Phys. 23, 3-11.

Harley, N.H. and Pasternack, B.S., 1972a, "Experimental Absorption Measurements Applied to Lung Dose from Radon Daughters," Health Phys. 23, 771-782.

Harley, N.H. and Pasternack, B.S., 1982, "Environmental radon daughter alpha dose factors in a five-lobed human lung", Health Phys. 42, 789-799.

Heyder, J. and Gebhart, J., 1977, 'Gravitational Deposition of Particles from Laminar Aerosol Flows Through Inclined Circular Tubes," Aerosol Sci., 8, 289-295.

Hislop, A., Muir, D.C.F., Jacobson, M., Simon, G. and Reid, L., 1972. 'Postnatal growth and functions of the pre-acinar airways". Thorax. 27, 265-

Hofmann, W., 1982, "Personal characteristics and environmental factors influencing lung dosimetry of inhaled radon decay products". In "Radiation Hazards in Mining: Control, Measurement, and Medical Aspects". Ed. M. Gormez. Soc. of Mining Engineers, New York, NY pp 669-674.

Hofmann, W., 1982a, "Dose Calculations for the Respiratory Tract from Inhaled Natural Radioactive Nuclides as a Function of Age - II Basal Cell Dose Distributions and Associated Lung Cancer Risk," Health Phys. 43, 31-44.

Hofmann, W., Steinhäusler, F. and Pohl, E., 1979, "Dose Calculations for the Respiratory Tract from Inhaled Natural Radioactive Nuclides as a Function of Age - Part I: Compartmental Deposition, Retention and Resulting Dose," Health Phys. 37, 517-523.

Hofmann, W., Steinhäusler, F and Pohl, E., 1980, 'Age-, sex- and weight-dependent dose patterns due to inhaled natural radionuclides". Proc. Symp. on "Natural Radiation Environment III", Houston, USA. CONF-780422-Vol 2. DOE-Symp. Ser. 51 pp 1116-1143.

Holaday, D.A., 1969, "History of exposure of miners to radon". Health Phys. 16, 547-552.

Horsfield, K., Relea, F.G. and Cumming, G., 1976, "Diameter, Length and Branching Ratios in the Bronchial Tree," Resp. Physiol. 26, 351-356.

Hursh, J.B. and Mercer, T.T., 1970, "Measurement of $^{212}Pb$ Loss Rate from Human Lungs," J. Appl. Physiol, 28, 268-274.

Hursh, J.B., Schraub, A., Sattler, E.L. and Hofmann, H.P., 1969, "Fate of $^{212}Pb$ Inhaled by Human Subjects," Health Phys. 16, 257-267.

HWC, 1980, "Guidelines for bioassay programs". 80-EHD-56, Health and Welfare Canada, Ottawa, Ontario.

IAEA, 1976, "Manual on radiological safety in uranium and thorium mines and mills". Safety Series No. 43. International Atomic Energy Agency. Vienna.

IAEA, 1980, "Basic requirements for personnel monitoring". Safety Series No. 14. International Atomic Energy Agency. Vienna.

ICRP, 1959, "Permissible dose for internal radiation". ICRP Publication 2. International Commission on Radiological Protection. Pergamon, Oxford.

ICRP, 1967, "Evaluation of radiation doses to body tissues from internal contamination due to occupational exposure". ICRP Publ. 10. International Commission on Radiological Protection, Pergamon, Oxford.

ICRP, 1968, "General principles of monitoring for radiation protection of workers". ICRP Publ. 12. International Commission on Radiological Protection, Pergamon, Oxford.

ICRP, 1975, "Task Group on Reference Man," ICRP Publ. 23, International Commission on Radiological Protection. Pergamon, Oxford.

ICRP, 1977, "Recommendations of the International Commission on Radiological Protection," ICRP Publ. 26, Ann. of ICRP, 1 (3), International Commission on Radiological Protection. Pergamon, Oxford.

ICRP, 1977a, "Radiation protection in uranium and other mines". ICRP Publ. 24. International Commission on Radiological Protection, Ann. of ICRP, 1 (1), Pergamon, Oxford.

ICRP, 1979, "ICRP Committee 2: Limits on Intakes of Radionuclides by Workers," ICRP Publ. 30, Part 1, Ann. of ICRP, 2 (3/4), International Commission on Radiological Protection. Pergamon, Oxford.

ICRP, 1980, "Biological effects of inhaled radionuclides". ICRP Publ. 31. International Commission on Radiological Protection, Ann. of ICRP, 4 (1/2), Pergamon, Oxford.

ICRP, 1980a, "Statement and recommendations of the 1980 Brighton meeting of the ICRP. "International Commission on Radiological Protection, Ann. of ICRP, 4 (2), Pergamon, Oxford.

ICRP, 1981, "Limits for inhalation of radon daughters by workers". ICRP Publ. 32. International Commission on Radiological Protection, Ann. of ICRP, 6 (1), Pergamon, Oxford.

Jackson, P.O., Cooper, J.A., Langford, J.C and Petersen, M.R., 1982, "Characteristics of attached radon-222 daughters under both laboratory and underground uranium mine environments". In "Radiation Hazards in Mining: Control, Measurement, and Medical Aspects". Ed. M. Gomez. Soc. of Mining Engineers, New York, NY, pp 1031-1042.

Jacobi, W., Aurand, K. and Schraub, A., 1957, "The Radiation Exposure of the Organism by Inhalation of Naturally Occurring Radioactive Aerosols," in "Advances in Radiobiology," Oliver and Boyd, Edinburgh, pp 310-318.

Jacobi, W., 1964, "The dose to the human respiratory tract by inhalation of short-lived $^{222}$Rn- and $^{220}$Rn- decay products". Health Phys. 10, 1163-1174.

Jacobi, W., 1972, "Activity and Potential $\alpha$-Energy of $^{222}$Rn and $^{220}$Rn-daughters in Different Air Atmospheres," Health Phys. 22, 441-450.

Jacobi, W., 1977, "Interpretation of Measurements in Uranium Mines: Dose Evaluation and Biomedical Aspects," in "Personal Dosimetry and Area Monitoring Suitable for Radon and Daughter Products," Proc. NEA Specialist Meeting, Elliot Lake, Canada, 4-8 October, 1976, OECD-NEA, pp 33-48.

Jacobi, W. and Eisfeld, K., 1980, "Dose to Tissues and Effective Dose Equivalent by Inhalation of Radon-222, Radon-220 and their Short-lived Daughters," GSF Report S-626.

Jacobi, W., 1981, "Indoor and outdoor exposure to short-lived radon-222 daughters: Dose factors and mean annual dose". Report to NEA Expert Group on Radon Dosimetry and Monitoring (unpublished).

Jacobi, W. and Eisfeld, K., 1981, "Internal Dosimetry of Radon-222, Radon-220 and their Short-lived Daughters," Proc. 2nd Special Symp. on "Natural Radiation Environment," Jan. 19-23, Bhabha Atomic Research Centre, Bombay (in press).

James, A.C., 1977, "Bronchial Deposition of Free Ions and Submicron Particles Studies in Excised Lung," Inhaled Particles IV (Ed. W.H. Walton), Pergamon Press, Oxford, pp 203-218.

James, A.C., Greenhalgh, J.R. and Smith, H., 1977, "Clearance of Lead-212 Ions from Rabbit Bronchial Epithelium to Blood," Phys. Med. Biol. 22, 932-948.

James, A.C., 1980, "Dosimetry aspects of radon, thoron and their daughters". Report to NEA Expert Group on Radon Dosimetry and Monitoring (unpublished).

James, A.C., Greenhalgh, J.R. and Birchall, A., 1980, "A Dosimetric Model for Tissues of the Human Respiratory Tract at Risk from Inhaled Radon and Thoron Daughters," in "Radiation Protection. A Systematic Approach to Safety," Proc. 5th Congress IRPA, Jerusalem, March 1980. Vol. 2, Pergamon Press, Oxford pp 1045-1048.

James, A.C., 1982, "Relative hazard of radon daughter exposure in mines and homes: The dosimetric approach". In "Radiological Protection - Advances in Theory and Practice". Proc. 3rd Int. Symp. SRP. Inverness, Scotland. Vol. 1, pp 411-417.

James, A.C., Jacobi, W and Steinhausler, F., 1982, "Respiratory tract dosimetry of radon and thoron daughters: The state-of-the-art and implications for epidemiology and radiobiology". In "Radiation Hazards in Mining: Control, Measurement and Medical Aspects". Ed. M. Gomez. Soc. of Mining Engineers, New York, NY, pp 42-54.

Johnson, J.R., 1975, "Estimation, recording and reporting of whole body doses from tritium oxide exposure at CRNL". AECL-5507, Atomic Energy of Canada Ltd.

Johnson, J.R., 1978, "Uncertainties in estimating working level months". AECL-6402, Atomic Energy of Canada Ltd.

Johnson, J.R. and Dunford, D.W., 1982, "Dose conversion factors for intake of selected radionuclides by infants and adults". AECL-7919, Atomic Energy of Canada Ltd.

Johnson, J.R. and Leach, V.A., 1982, "An examination of the relationship between WLM exposure and dose". In "Radiation Hazards in Mining: Control, Measurement and Medical Aspects". Ed. M. Gomez. Soc. of Mining Engineers, New York, NY, pp 390-397.

Keller, G., Folkerts, K.H. and Muth, H., 1982, "Measurements of the activity concentrations of $^{222}Rn$ and its decay products in German houses, dose calculations and estimate of risk". In "Radiological Protection - Advances in Theory and Practice". Proc. 3rd Int. Symp. SRP Inverness, Scotland, Vol.1, pp.267-274.

Kunz, E., Sevc, J., Placek, V. and Horacek, J., 1979, "Lung Cancer in Man in Relation to Different Time Distribution of Radiation Exposure," Health Phys. 36, 699-706.

Leach, V.A. and Lokan, K.H., 1979, "Monitoring employee exposure to radon and its daughters in uranium mines". Australian Radiation Laboratory. ARL/TR 011.

Martin, D. and Jacobi, W., 1972, "Diffusion Deposition of Small-sized Particles in the Bronchial Tree," Health Phys. 23, 23-29.

Martonen, T.B. and Patel, M., 1981, "Modelling the dose distribution of $H_2SO_4$ aerosols in the human tracheobronchial tree". Am. Ind. Hyg. Assoc. J. 42, 453-460.

McGregor, R.G., Vasudev, P., Letourneau, E.G., et al., 1980, "Background concentrations of radon and radon daughters in Canadian homes". Health Phys. 39, 285-289.

Mercer, T.T. and Stowe, W.A., 1977, "Radioactive Aerosols Produced by Radon in Room Air," Inhaled Particles III (Ed. W.H. Walton). Unwin Bros., Old Woking, pp 839-851.

Mercer, T.T., 1975, "Unattached Radon Decay products in Mine Air," Health Phys. 28, 158-161.

Mohnen, V., 1967, "Investigation of the Attachment of Neutral and Electrically Charged Emanation Decay Products to Aerosols," Doctoral Thesis, AERE Transl. 1106.

OECD, 1979, "Exposure to radiation from the natural radioactivity in building materials". Report by an Expert Group. Nuclear Energy Agency, Paris.

Orzechowski, W., Cruscielewski, W. and Domanski, T., 1981, "Some features of Kodak LR-115 Type II foil for its application in the measurement of exposure to radon and its progeny". Proc. Specialist Meeting "The Assessment of Radon and Daughter Exposure and Related Biological Effects". Rome, Italy, RD Press, Univ. pp 20-29.

Pohl, E., 1962, "Die Strahlenbelastung bei der Inhalation von Radium-Emanation". Strahlentherapie, 119, 77-96.

Postendorfer, J. and Mercer, T.T., 1978, "Influence of Nuclei Concentration and Humidity upon the Attachment Rate of Atoms in the Atmosphere," Atmos. Envir., 12, 2223-2228.

Rafferty, P.J. and Steinhäusler, F., 1982, "The OECD-Nuclear Energy Agency Programme on Dosimetry and Monitoring of Radon, Thoron and Their Decay Products". In "Radiation Hazards in Mining: Control, Measurement and Medical Aspects". Ed. M. Gomez. Soc. of Mining Engineers, New York, NY pp 17-22.

Saccomanno, G., Archer, V.E., Auerbach, O., et al, 1982, "Age factor in histological type of lung cancer among uranium miners: A preliminary report". In "Radiation Hazards in Mining: Control, Measurement and Medical Aspects". Ed. M. Gomez. Soc. of Mining Engineers, New York, NY pp 675-679.

Schlesinger, R.B and Lippmann, M., 1978, 'Selective Particle Deposition and Bronchogenic Carcinoma," Envir. Res. 15, 424-431.

Sturgess, J.M., 1977, "Structural Organisation of Mucus in the Lung," in "Pulmonary Macrophage and Endotheial Cells," (Eds. C.L. Sanders et al) ERDA Symp. Ser. 43, CONF-760927, pp 149-161.

UNSCEAR, 1977, United Nations Scientific Committee on the Effects of Atomic Radiation. Report to the General Assembly, Annexes G & B, UN, New York.

UNSCEAR, 1982, "United Nations Scientific Committee on the Effects of Atomic Radiation. Report to the General Assembly. Annex. UN, New York (in press).

van der Vooren, A.W. and Phillips, C.R., 1982, "A comparison of analytical models for diffusional deposition of radon daughters in a human lung bifurcation". In "Radiation Hazards in Mining: Control, Measurement, and medical Aspects". Ed. M. Gomez. Soc. of Mining Engineers, New York, NY pp 55-62.

Walsh, P.J., 1970, "Radiation Dose to the Respiratory Tract of Uranium Miners -A Review of the Literature," Envir. Res. 3, 14-36.

Walsh, P.J. and Hamrick, P.E., 1977, "Radioactive materials - determinants of dose to the respiratory tract". In "Handbook of Physiology Section 9: Reactions to Environmental Agents". Ed. D.H.K. Lee. Am. Physiol Soc. Bethesda, Md. pp 233-242.

Weibel, E.R., 1963. "Morphometry of the Human Lung", Springer-Verlag, Berlin.

Wise, K.N., 1982, "Dose Conversion Factors for Radon Daughters in Underground and Open-cut Mine Atmospheres," Health Phys 43, 53-64.

Yeates, D.B., Gerrity, T.R. and Garrard, C.S., 1981, "Characteristics of Tracheobronchial Deposition and Clearance in Man," Inhaled Particles V (Ed. W.H. Walton), Pergamon, Oxford, pp 245-258.

Yeh, H.C., 1974, "Use of a Heat Transfer Analogy for a Mathematical Model of Repiratory Tract Deposition," Bull. Math. Biol. 36, 105-116.

Yeh, H.C. and Schum, G.M., 1980, "Models of the Human Lung Airways and their Application to Inhaled Particle Deposition," Bull. Math. Biol. 42, 461-480.

Zettwoog, P., 1981, "The alpha-individual dosimetry in French uranium mines". Commissariat à l'Energie Atomique. SPT Internal Report No. 234.

# OTHER NEA PUBLICATIONS

## RADIATION PROTECTION

Recommendations for Ionization Chamber Smoke Detectors in Implementation of Radiation Protection Standards (1977)

## RADIOPROTECTION

Recommandations relatives aux détecteurs de fumée à chambre d'ionisation en application des normes de radioprotection (1977)

Free on request – Gratuit sur demande

Radon Monitoring
(Proceedings of the NEA Specialist Meeting, Paris, 1978)

Surveillance du radon
(Compte rendu d'une réunion de spécialistes de l'AEN, Paris, 1978)

£8.00 US$16.50 F66,00

Management, Stabilisation and Environmental Impact of Uranium Mill Tailings
(Proceedings of the Albuquerque Seminar, United States, 1978)

Gestion, stabilisation et incidence sur l'environnement des résidus de traitement de l'uranium
(Compte rendu du Séminaire d'Albuquerque, États-Unis, 1978)

£9.80 US$20.00 F80,00

Exposure to Radiation from the Natural Radioactivity in Building Materials
(Report by an NEA Group of Experts, 1979)

Exposition aux rayonnements due à la radioactivité naturelle des matériaux de construction
(Rapport établi par un Groupe d'experts de l'AEN, 1979)

Free on request – Gratuit sur demande

Marine Radioecology
(Proceedings of the Tokyo Seminar, 1979)

Radioécologie marine
(Compte rendu du Colloque de Tokyo, 1979)

£9.60 US$21.50 F86,00

Radiological Significance and Management of Tritium, Carbon-14, Krypton-85 and Iodine-129 arising from the Nuclear Fuel Cycle
(Report by an NEA Group of Experts, 1980)

Importance radiologique et gestion des radionucléides : tritium, carbone-14, krypton-85 et iode-129, produits au cours du cycle du combustible nucléaire
(Rapport établi par un Groupe d'experts de l'AEN, 1980)

£8.40 US$19.00 F76,00

The Environmental and Biological Behaviour of Plutonium and Some Other Transuranium Elements
(Report by an NEA Group of Experts, 1981)

Le comportement mésologique et biologique du plutonium et de certains autres éléments transuraniens
(Rapport établi par un Groupe d'experts de l'AEN, 1981)

£4.60 US$10.00 F46,00

Uranium Mill Tailing Management
(Proceedings of two Workshops Fort Collins, USA, 1981)

La gestion des résidus de traitement de l'uranium
(Compte rendu de deux réunions de travail Fort Collins, États-Unis, 1981)

£7.20 US$16.00 F72,00

## RADIOACTIVE WASTE MANAGEMENT

## GESTION DES DÉCHETS RADIOACTIFS

Objectives, Concepts and Strategies for the Management of Radioactive Waste Arising from Nuclear Power Programmes (Report by an NEA Group of Experts, 1977)

Objectifs, concepts et stratégies en matière de gestion des déchets radioactifs résultant des programmes nucléaires de puissance (Rapport établi par un Groupe d'experts de l'AEN, 1977)

£8.50 US$17.50 F70,00

Decision of the OECD Council of the 22nd of July 1977 establishing a Multilateral Consultation and Surveillance Mechanism for Sea Dumping of Radioactive Waste

Décision du Conseil de l'OCDE en date du 22 juillet 1977 instituant un Mécanisme multilatéral de consultation et de surveillance pour l'immersion de déchets radioactifs en mer

Free on request – Gratuit sur demande

In Situ Heating Experiments in Geological Formations (Proceedings of the Ludvika Seminar, Sweden, 1978)

Expériences de dégagement de chaleur in situ dans les formations géologiques (Compte rendu du Séminaire de Ludvika, Suède, 1978)

£8.00 US$16.50 F66,00

Migration of Long-lived Radionuclides in the Geosphere (Proceedings of the Brussels Workshop, 1979)

Migration des radionucléides à vie longue dans la géosphère (Compte rendu de la réunion de travail de Bruxelles, 1979)

£8.30 US$17.00 F68,00

Low-Flow, Low-Permeability Measurements in Largely Impermeable Rocks (Proceedings of the Paris Workshop, 1979)

Mesures des faibles écoulements et des faibles perméabilités dans des roches relativement imperméables (Compte rendu de la réunion de travail de Paris, 1979)

£7.80 US$16.00 F64,00

On-Site Management of Power Reactor Wastes (Proceedings of the Zurich Symposium, 1979)

Gestion des déchets en provenance des réacteurs de puissance sur le site de la centrale (Compte rendu du Colloque de Zurich, 1979)

£11.00 US$22.50 F90,00

Recommended Operational Procedures for Sea Dumping of Radioactive Waste (1979)

Recommandations relatives aux procédures d'exécution des opérations d'immersion de déchets radioactifs en mer (1979)

Free on request – Gratuit sur demande

Guidelines for Sea Dumping Packages of Radioactive Waste (Revised version, 1979)

Guide relatif aux conteneurs de déchets radioactifs destinés au rejet en mer (Version révisée, 1979)

Free on request – Gratuit sur demande

Use of Argillaceous Materials for the Isolation of Radioactive Waste
(Proceedings of the Paris Workshop, 1979)

Utilisation des matériaux argileux pour l'isolement des déchets radioactifs
(Compte rendu de la réunion de travail de Paris, 1979)

£7.60 US$17.00 F68,00

Review of the Continued Suitability of the Dumping Site for Radioactive Waste in the North-East Atlantic (1980)

Réévaluation de la validité du site d'immersion de déchets radioactifs dans la région nord-est de l'Atlantique (1980)

Free on request – Gratuit sur demande

Decommissioning Requirements in the Design of Nuclear Facilities
(Proceedings of the NEA Specialist Meeting, Paris, 1980)

Déclassement des installations nucléaires : exigences à prendre en compte au stade de la conception
(Compte rendu d'une réunion de spécialistes de l'AEN, Paris, 1980))

£7.80 US$17.50 F70,00

Borehole and Shaft Plugging
(Proceedings of the Columbus Workshop, United States, 1980)

Colmatage des forages et des puits
(Compte rendu de la réunion de travail de Columbus, États-Unis, 1980)

£12.00 US$30.00 F120,00

Radionuclide Release Scenarios for Geologic Repositories
(Proceedings of the Paris Workshop, 1980)

Scénarios de libération des radionucléides à partir de dépôts situés dans les formations géologiques
(Compte rendu de la réunion de travail de Paris, 1980)

£6.00 US$15.00 F60,00

Research and Environmental Surveillance Programme Related to Sea Disposal of Radioactive Waste (1981)

Programme de recherches et de surveillance du milieu lié à l'immersion de déchets radioactifs en mer (1981)

Free on request – Gratuit sur demande

Cutting Techniques as related to Decommissioning of Nuclear Facilities
(Report by an NEA Group of Experts, 1981)

Techniques de découpe utilisées au cours du déclassement d'installations nucléaires
(Rapport établi par un Groupe d'experts de l'AEN, 1981)

£3.00 US$7.50 F30,00

Decontamination Methods as related to Decommissioning of Nuclear Facilities
(Report by an NEA Group of Experts, 1981)

Méthodes de décontamination relatives au déclassement des installations nucléaires
(Rapport établi par un Groupe d'experts de l'AEN, 1981)

£2.80 US$7.00 F28,00

Siting of Radioactive Waste Repositories in Geological Formations
(Proceedings of the Paris Workshop, 1981)

Choix des sites des dépôts de déchets radioactifs dans les formations géologiques
(Compte rendu d'une réunion de travail de Paris, 1981)

£6.80 US$15.00 F68,00

Near-Field Phenomena in Geologic Repositories for Radioactive Waste
(Proceedings of the Seattle Workshop, United States, 1981)

Phénomènes en champ proche des dépôts de déchets radioactifs en formations géologiques
(Compte rendu de la réunion de travail de Seattle, États-Unis, 1981)

£11.00 US$24.50 F110,00

Disposal of Radioactive Waste – An Overview of the Principles Involved, 1982

Évacuation des déchets radioactifs – un aperçu des principes en vigueur, 1982

Free on request – Gratuit sur demande

Geological Disposal of Radioactive Waste – *Research in the OECD Area* (1982)

Évacuation des déchets radioactifs dans les formations géologiques – *Recherches effectuées dans les pays de l'OCDE* (1982)

Free on request – Gratuit sur demande

Geological Disposal of Radioactive Waste: *Geochemical Processes* (1982)

Évacuation des déchets radioactifs dans les formations géologiques : *Processus géochimique* (1982)

£7.00 US$14.00 F70.00

Geological Disposal of Radioactive Waste: *In Situ Experiments in Granite*
(Proceedings of the Stockholm Workshop, 1982)

Évacuation des déchets radioactifs dans les formations géologiques : *Expériences in situ dans du granite*
(Compte rendu d'une réunion de travail de Stokholm, 1982)

£10.00 US$20.00 F100,00

Interim Oceanographic Description of the North-East Atlantic Site for the Disposal of Low-Level Radioactive Waste (1983)

État des connaissances océanographiques relatives au site d'immersion de déchets radioactifs de faible activité dans l'Atlantique nord-est (1983)

Free on request – Gratuit sur demande

The International Stripa Project: Background and Research Results (1983)

Projet international de Stripa : Informations générales et résultats des recherches (1983)

Free on request – Gratuit sur demande

International Co-operation for Safe Radioactive Waste Management (1983)

Une coopération internationale pour une gestion sûre des déchets radioactifs (1983)

Free on request – Gratuit sur demande

Long Term Management of High Level Radioactive Waste – The Meaning of a Demonstration

Gestion à long terme des déchets de haute activité – Signification d'une démonstration

Free on request – Gratuit sur demande

• • •

# OECD SALES AGENTS
# DÉPOSITAIRES DES PUBLICATIONS DE L'OCDE

**ARGENTINA – ARGENTINE**
Carlos Hirsch S.R.L., Florida 165, 4° Piso (Galería Guemes)
1333 BUENOS AIRES, Tel. 33.1787.2391 y 30.7122

**AUSTRALIA – AUSTRALIE**
Australia and New Zealand Book Company Pty. Ltd.,
10 Aquatic Drive, Frenchs Forest, N.S.W. 2086
P.O. Box 459, BROOKVALE, N.S.W. 2100

**AUSTRIA – AUTRICHE**
OECD Publications and Information Center
4 Simrockstrasse 5300 BONN. Tel. (0228) 21.60.45
Local Agent/Agent local :
Gerold and Co., Graben 31, WIEN 1. Tel. 52.22.35

**BELGIUM – BELGIQUE**
Jean De Lannoy, Service Publications OCDE
avenue du Roi 202, B-1060 BRUXELLES. Tel. 02/538.51.69

**BRAZIL – BRÉSIL**
Mestre Jou S.A., Rua Guaipa 518,
Caixa Postal 24090, 05089 SAO PAULO 10. Tel. 261.1920
Rua Senador Dantas 19 s/205-6, RIO DE JANEIRO GB.
Tel. 232.07.32

**CANADA**
Renouf Publishing Company Limited,
2182 ouest, rue Ste-Catherine,
MONTRÉAL, Qué. H3H 1M7. Tel. (514)937.3519
OTTAWA, Ont. K1P 5A6, 61 Sparks Street

**DENMARK – DANEMARK**
Munksgaard Export and Subscription Service
35, Nørre Søgade
DK 1370 KØBENHAVN K. Tel. +45.1.12.85.70

**FINLAND – FINLANDE**
Akateeminen Kirjakauppa
Keskuskatu 1, 00100 HELSINKI 10. Tel. 65.11.22

**FRANCE**
Bureau des Publications de l'OCDE,
2 rue André-Pascal, 75775 PARIS CEDEX 16. Tel. (1) 524.81.67
Principal correspondant :
13602 AIX-EN-PROVENCE : Librairie de l'Université.
Tel. 26.18.08

**GERMANY – ALLEMAGNE**
OECD Publications and Information Center
4 Simrockstrasse 5300 BONN Tel. (0228) 21.60.45

**GREECE – GRÈCE**
Librairie Kauffmann, 28 rue du Stade,
ATHÈNES 132. Tel. 322.21.60

**HONG-KONG**
Government Information Services,
Publications/Sales Section, Baskerville House,
2/F., 22 Ice House Street

**ICELAND – ISLANDE**
Snaebjörn Jönsson and Co., h.f.,
Hafnarstraeti 4 and 9, P.O.B. 1131, REYKJAVIK.
Tel. 13133/14281/11936

**INDIA – INDE**
Oxford Book and Stationery Co. :
NEW DELHI-1, Scindia House. Tel. 45896
CALCUTTA 700016, 17 Park Street. Tel. 240832

**INDONESIA – INDONÉSIE**
PDIN-LIPI, P.O. Box 3065/JKT., JAKARTA, Tel. 583467

**IRELAND – IRLANDE**
TDC Publishers – Library Suppliers
12 North Frederick Street, DUBLIN 1 Tel. 744835-749677

**ITALY – ITALIE**
Libreria Commissionaria Sansoni :
Via Lamarmora 45, 50121 FIRENZE. Tel. 579751/584468
Via Bartolini 29, 20155 MILANO. Tel. 365083
Sub-depositari :
Ugo Tassi
Via A. Farnese 28, 00192 ROMA. Tel. 310590
Editrice e Libreria Herder,
Piazza Montecitorio 120, 00186 ROMA. Tel. 6794628
Costantino Ercolano, Via Generale Orsini 46, 80132 NAPOLI. Tel. 405210
Libreria Hoepli, Via Hoepli 5, 20121 MILANO. Tel. 865446
Libreria Scientifica, Dott. Lucio de Biasio "Aeiou"
Via Meravigli 16, 20123 MILANO Tel. 807679
Libreria Zanichelli
Piazza Galvani 1/A, 40124 Bologna Tel. 237389
Libreria Lattes, Via Garibaldi 3, 10122 TORINO. Tel. 519274
La diffusione delle edizioni OCSE è inoltre assicurata dalle migliori librerie nelle città più importanti.

**JAPAN – JAPON**
OECD Publications and Information Center,
Landic Akasaka Bldg., 2-3-4 Akasaka,
Minato-ku, TOKYO 107 Tel. 586.2016

**KOREA – CORÉE**
Pan Korea Book Corporation,
P.O. Box n° 101 Kwangwhamun, SÉOUL. Tel. 72.7369

**LEBANON – LIBAN**
Documenta Scientifica/Redico,
Edison Building, Bliss Street, P.O. Box 5641, BEIRUT.
Tel. 354429 – 344425

**MALAYSIA – MALAISIE**
University of Malaya Co-operative Bookshop Ltd.
P.O. Box 1127, Jalan Pantai Baru
KUALA LUMPUR. Tel. 51425, 54058, 54361

**THE NETHERLANDS – PAYS-BAS**
Staatsuitgeverij, Verzendboekhandel,
Chr. Plantijnstraat 1 Postbus 20014
2500 EA S-GRAVENHAGE. Tel. nr. 070.789911
Voor bestellingen: Tel. 070.789208

**NEW ZEALAND – NOUVELLE-ZÉLANDE**
Publications Section,
Government Printing Office Bookshops:
AUCKLAND: Retail Bookshop: 25 Rutland Street,
Mail Orders: 85 Beach Road, Private Bag C.P.O.
HAMILTON: Retail Ward Street,
Mail Orders, P.O. Box 857
WELLINGTON: Retail: Mulgrave Street (Head Office),
Cubacade World Trade Centre
Mail Orders: Private Bag
CHRISTCHURCH: Retail: 159 Hereford Street,
Mail Orders: Private Bag
DUNEDIN: Retail: Princes Street
Mail Order: P.O. Box 1104

**NORWAY – NORVÈGE**
J.G. TANUM A/S Karl Johansgate 43
P.O. Box 1177 Sentrum OSLO 1. Tel. (02) 80.12.60

**PAKISTAN**
Mirza Book Agency, 65 Shahrah Quaid-E-Azam, LAHORE 3.
Tel. 66839

**PHILIPPINES**
National Book Store, Inc.
Library Services Division, P.O. Box 1934, MANILA.
Tel. Nos. 49.43.06 to 09, 40.53.45, 49.45.12

**PORTUGAL**
Livraria Portugal, Rua do Carmo 70-74,
1117 LISBOA CODEX. Tel. 360582/3

**SINGAPORE – SINGAPOUR**
Information Publications Pte Ltd,
Pei-Fu Industrial Building,
24 New Industrial Road N° 02-06
SINGAPORE 1953, Tel. 2831786, 2831798

**SPAIN – ESPAGNE**
Mundi-Prensa Libros, S.A.
Castelló 37, Apartado 1223, MADRID-1. Tel. 275.46.55
Libreria Bosch, Ronda Universidad 11, BARCELONA 7.
Tel. 317.53.08, 317.53.58

**SWEDEN – SUÈDE**
AB CE Fritzes Kungl Hovbokhandel,
Box 16 356, S 103 27 STH, Regeringsgatan 12,
DS STOCKHOLM. Tel. 08/23.89.00
Subscription Agency/Abonnements:
Wennergren-Williams AB,
Box 13004, S104 25 STOCKHOLM.
Tel. 08/54.12.00

**SWITZERLAND – SUISSE**
OECD Publications and Information Center
4 Simrockstrasse 5300 BONN. Tel. (0228) 21.60.45
Local Agents/Agents locaux
Librairie Payot, 6 rue Grenus, 1211 GENÈVE 11. Tel. 022.31.89.50

**TAIWAN – FORMOSE**
Good Faith Worldwide Int'l Co., Ltd.
9th floor, No. 118, Sec. 2,
Chung Hsiao E. Road
TAIPEI. Tel. 391.7396/391.7397

**THAILAND – THAILANDE**
Suksit Siam Co., Ltd., 1715 Rama IV Rd.
Samyan, BANGKOK 5. Tel. 2511630

**TURKEY – TURQUIE**
Kültur Yayinlari Is-Türk Ltd. Sti.
Atatürk Bulvari No : 77/B
KIZILAY/ANKARA. Tel. 17 02 66
Dolmabahce Cad. No : 29
BESIKTAS/ISTANBUL. Tel. 60 71 88

**UNITED KINGDOM – ROYAUME-UNI**
H.M. Stationery Office, P.O.B. 276,
LONDON SW8 5DT. Tel. (01) 622.3316, or
49 High Holborn, LONDON WC1V 6 HB (personal callers)
Branches at: EDINBURGH, BIRMINGHAM, BRISTOL,
MANCHESTER, BELFAST.

**UNITED STATES OF AMERICA – ÉTATS-UNIS**
OECD Publications and Information Center, Suite 1207,
1750 Pennsylvania Ave., N.W. WASHINGTON, D.C.20006 – 4582
Tel. (202) 724.1857

**VENEZUELA**
Libreria del Este, Avda. F. Miranda 52, Edificio Galipan,
CARACAS 106. Tel. 32.23.01/33.26.04/31.58.38

**YUGOSLAVIA – YOUGOSLAVIE**
Jugoslovenska Knjiga, Knez Mihajlova 2, P.O.B. 36, BEOGRAD.
Tel. 621.992

Les commandes provenant de pays où l'OCDE n'a pas encore désigné de dépositaire peuvent être adressées à :
OCDE, Bureau des Publications, 2, rue André-Pascal, 75775 PARIS CEDEX 16.

Orders and inquiries from countries where sales agents have not yet been appointed may be sent to:
OECD, Publications Office, 2, rue André-Pascal, 75775 PARIS CEDEX 16.

67048-10-1983

OECD PUBLICATIONS, 2, rue André-Pascal, 75775 PARIS CEDEX 16 - No. 42721 1983
PRINTED IN FRANCE
(66 83 06 1) ISBN 92-64-12520-5